Lecture Notes in Physics

For information about Vols. 1–44, please contact your bookseller or Springer-Verlag.

Vol. 45: Dynamical Concepts on Scaling Violation and the New Resonances in e^+e^- Annihilation. Edited by B. Humpert. VII, 248 pages. 1976.

Vol. 46: E. J. Flaherty, Hermitian and Kählerian Geometry in Relativity. VIII, 365 pages. 1976.

Vol. 47: Padé Approximants Method and Its Applications to Mechanics. Edited by H. Cabannes. XV, 267 pages. 1976.

Vol. 48: Interplanetary Dust and Zodiacal Light. Proceedings 1975. Edited by H. Elsässer and H. Fechtig. XII, 496 pages. 1976.

Vol. 49: W. G. Harter and C. W. Patterson, A Unitary Calculus for Electronic Orbitals. XII, 144 pages. 1976.

Vol. 50: Group Theoretical Methods in Physics. 4th International Colloquium. Nijmegen 1975. Edited by A. Janner, T. Janssen, and M. Boon. XIII, 629 pages. 1976.

Vol. 51: W. Nörenberg and H. A. Weidenmüller. Introduction of the Theory of Heavy-Ion Collisions. 2nd enlarged edition. IX, 334 pages. 1980.

Vol. 52: M. Mladjenović, Development of Magnetic β-Ray Spectroscopy. X, 282 pages. 1976.

Vol. 53: D. J. Simms and N. M. J. Woodhouse, Lectures on Geometric Quantization. V, 166 pages. 1976.

Vol. 54: Critical Phenomena. Sitges International School on Statistical Mechanics, June 1976. Edited by J. Brey and R. B. Jones. XI, 383 pages. 1976.

Vol. 55: Nuclear Optical Model Potential. Proceedings 1976. Edited by S. Boffi and G. Passatore. VI, 221 pages. 1976.

Vol. 56: Current Induced Reactions. International Summer Institute, Hamburg 1975. Edited by J. G. Körner, G. Kramer, and D. Schildknecht. V, 553 pages. 1976.

Vol. 57: Physics of Highly Excited States in Solids. Proceedings 1975. Edited by M. Ueta and Y. Nishina. IX, 391 pages. 1976.

Vol. 58: Computing Methods in Applied Sciences. Proceedings 1975. Edited by R. Glowinski and J. L. Lions. VIII, 593 pages. 1976.

Vol. 59: Proceedings of the Fifth International Conference on Numerical Methods in Fluid Dynamics. 1976. Edited by A. I. van de Vooren and P. J. Zandbergen. VII, 459 pages. 1976.

Vol. 60: C. Gruber, A. Hintermann, and D. Merlini, Group Analysis of Classical Lattice Systems. XIV, 326 pages. 1977.

Vol. 61: International School on Electro and Photonuclear Reactions I. Edited by C. Schaerf. VIII, 650 pages. 1977.

Vol. 62: International School on Electro and Photonuclear Reactions II. Edited by C. Schaerf. VIII, 301 pages. 1977.

Vol. 63: V. K. Dobrev et al., Harmonic Analysis on the n-Dimensional Lorentz Group and Its Application to Conformal Quantum Field Theory. X, 280 pages. 1977.

Vol. 64: Waves on Water of Variable Depth. Edited by D. G. Provis and R. Radok. 231 pages. 1977.

Vol. 65: Organic Conductors and Semiconductors. Proceedings 1976. Edited by L. Pál, G. Grüner, A. Jánossy and J. Sólyom. 654 pages. 1977.

Vol. 66: A. H. Völkel, Fields, Particles and Currents. VI, 354 pages. 1977.

Vol. 67: W. Drechsler and M. E. Mayer, Fiber Bundle Techniques in Gauge Theories. X, 248 pages. 1977.

Vol. 68: Y. V. Venkatesh, Energy Methods in Time-Varying System Stability and Instability Analyses. XII, 256 pages. 1977.

Vol. 69: K. Rohlfs, Lectures on Density Wave Theory. VI, 184 pages. 1977.

Vol. 70: Wave Propagation and Underwater Acoustics. Edited by J. Keller and J. Papadakis. VIII. 287 pages. 1977.

Vol. 71: Problems of Stellar Convection. Proceedings 1976. Edited by E. A. Spiegel and J. P. Zahn. VIII, 363 pages. 1977.

Vol. 72: Les instabilités hydrodynamiques en convection libre forcée et mixte. Edité par J. C. Legros et J. K. Platten. X, 202 pages. 1978.

Vol. 73: Invariant Wave Equations. Proceedings 1977. Edited by G. Velo and A. S. Wightman. VI, 416 pages. 1978.

Vol. 74: P. Collet and J.-P. Eckmann, A Renormalization Group Analysis of the Hierarchical Model in Statistical Mechanics. IV, 199 pages. 1978.

Vol. 75: Structure and Mechanisms of Turbulence I. Proceedings 1977. Edited by H. Fiedler. XX, 295 pages. 1978.

Vol. 76: Structure and Mechanisms of Turbulence II. Proceedings 1977. Edited by H. Fiedler. XX, 406 pages. 1978.

Vol. 77: Topics in Quantum Field Theory and Gauge Theories. Proceedings, Salamanca 1977. Edited by J. A. de Azcárraga. X, 378 pages 1978.

Vol. 78: Böhm, The Rigged Hilbert Space and Quantum Mechanics. IX, 70 pages. 1978.

Vol. 79: Group Theoretical Methods in Physics. Proceedings, 1977. Edited by P. Kramer and A. Rieckers. XVIII, 546 pages. 1978.

Vol. 80: Mathematical Problems in Theoretical Physics. Proceedings, 1977. Edited by G. Dell'Antonio, S. Doplicher and G. Jona-Lasinio. VI, 438 pages. 1978.

Vol. 81: MacGregor, The Nature of the Elementary Particle. XXII, 482 pages. 1978.

Vol. 82: Few Body Systems and Nuclear Forces I. Proceedings, 1978. Edited by H. Zingl, M. Haftel and H. Zankel. XIX, 442 pages. 1978.

Vol. 83: Experimental Methods in Heavy Ion Physics. Edited by K. Bethge. V, 251 pages. 1978.

Vol. 84: Stochastic Processes in Nonequilibrium Systems, Proceedings, 1978. Edited by L. Garrido, P. Seglar and P. J. Shepherd. XI, 355 pages. 1978

Vol. 85: Applied Inverse Problems. Edited by P. C. Sabatier. V, 425 pages. 1978.

Vol. 86: Few Body Systems and Electromagnetic Interaction. Proceedings 1978. Edited by C. Ciofi degli Atti and E. De Sanctis. VI, 352 pages. 1978.

Vol. 87: Few Body Systems and Nuclear Forces II, Proceedings, 1978. Edited by H. Zingl, M. Haftel, and H. Zankel. X, 545 pages. 1978.

Vol. 88: K. Hutter and A. A. F. van de Ven, Field Matter Interactions in Thermoelastic Solids. VIII, 231 pages. 1978.

Vol. 89: Microscopic Optical Potentials, Proceedings, 1978. Edited by H. V. von Geramb. XI, 481 pages. 1979.

Vol. 90: Sixth International Conference on Numerical Methods in Fluid Dynamics. Proceedings, 1978. Edited by H. Cabannes, M. Holt and V. Rusanov. VIII, 620 pages. 1979.

Lecture Notes in Physics

Edited by J. Ehlers, München, K. Hepp, Zürich
R. Kippenhahn, München, H. A. Weidenmüller, Heidelberg
and J. Zittartz, Köln
Managing Editor: W. Beiglböck, Heidelberg

129

Geometrical and Topological Methods in Gauge Theories

Proceedings of the Canadian Mathematical
Society Summer Research Institute
McGill University, Montréal
September 3 - 8, 1979

Edited by J. P. Harnad and S. Shnider

Springer-Verlag
Berlin Heidelberg GmbH 1980

Editors

John Harnad
Université de Montréal
Centre de Recherches de Mathématiques Appliquées
Case postale 6128, Succ. "A"
Montréal H3C 3J7
Canada

Steven Shnider
Department of Mathematics, McGill University
Montréal
Canada

ISBN 978-3-540-10010-2 ISBN 978-3-540-38142-6 (eBook)
DOI 10.1007/978-3-540-38142-6

Originally published by Springer-Verlag Berlin Heidelberg New York in 1980

2153/3140-543210

TABLE OF CONTENTS

CONTRIBUTED PAPERS

LIST OF PARTICIPANTS

INTRODUCTION

This volume is based on papers presented at the Canadian Mathematical Society Summer Research Institute on Geometrical and Topological Methods in Gauge Theories which took place at McGill University from September 3-8, 1979 supported by funds from the National Sciences and Engineering Research Council of Canada. During that period approximately fifty mathematicians and physicists exchanged ideas on a wide variety of topics related to gauge theory, which is becoming an increasingly popular field of interest for both groups. The fact that the variety of topics treated in the present volume actually involves a considerable overlap of concepts and methods demonstrates the unity that gauge theory has brought to seemingly unrelated branches of mathematics and physics. A number of experts active in research in these areas were present and shared their ideas in both formal lectures and informal conversation. The contents of the invited addresses included in this volume are summarized below.

C.Bernard discusses the validity of saddle point approximation in evaluating instanton contributions to the functional integral, and shows how to calculate the one-loop vacuum-vacuum amplitude about an instanton for the SU(N) gauge group. Then he gives arguments indicating that, in physical effects such as e^+e^- annihilation into hadrons at short distances or such as the proposed confining phase transition of Callen-Dashen-Gross, the important contributions are from instantons which are large in scale and for which the dilute gas approximation breaks down. Finally he discusses objections to instantons based on taking the large-N limit.

In the paper by E.Corrigan and P.Goddard it is shown that certain expressions which arise in approximating functional integrals by the method of stationary phase can be interpreted in terms of zeta functions for elliptic operators associated to the instanton fields. Standard index theory allows one to compute $\zeta(0)$, but the calculation of $\zeta'(0)$ is more difficult and is carried out indirectly. First, the authors find a variational formula for $\delta_{\zeta'}(0)$ which involves the Green's function of an instanton operator. The Atiyah-Drinfeld-Hitchen-Manin construction of instantons is used to give an explicit expression for the Green's function, but there remain unsolved problems in passing from $\delta_{\zeta'}(0)$ to $\zeta'(0)$. Finally a complete analysis of the two dimensional σ-model is discussed.

D.Fairlie surveys his own and other attempts to explain the Weinberg-Salam model in terms of a pure gauge theory in dimensions greater than 4, where Higgs fields appear after a dimensional reduction has distinguished certain components of the gauge fields. Various phenomenological problems which the different models encounter are discussed.

P.Freund discussed generalized spin manifolds which allow for the global definition of spinor fields possessing internal gauge symmetry when topology will not allow for the existence of ordinary global spinor fields. Through this, the calculation of the chiral Adler-Bell-Jackiw anomaly, which in some manifolds would seem to lead to fractional values for the helicity flip, is corrected to give even integral values. He also analyzes the particle spectrum in SU(8) models of quarks and leptons suggested by the non-linearly realized SU(8) gauge invariance of SO(8/4) supergravity.

R.Jackiw investigates Yang-Mills theory with sources finding, besides abelian coulomb solutions, certain non-abelian coulomb-like static solutions which tend to a pure gauge (not zero) as the source is turned off. These solutions are not known explicitly but are calculated perturbatively for weak sources. He also finds time dependent solutions generalizing each of the above types which decrease the energy of the corresponding static solution. Using a hamiltonian formulation to analyze stability as in classical mechanics, Jackiw shows that the above static solutions are stable without minimizing energy, due to "gyroscropic" cross terms of type $\vec{E} \times \vec{A}$ similar to $\vec{q} \times \vec{p}$ in the classical top. As source strength increases a bifurcation occurs, the abelian coulomb solution becomes unstable and numerical methods give only partial results on the other solutions.

L. O'Raifeartaigh shows how the long range forces between non-abelian monopoles may be computed analytically by making a suitable ansatz for the exterior boundary conditions. The cancellation between scalar attraction and magnetic repulsion known for spherically symmetric monopoles satisfying the Bogomolny equation is shown to no longer hold if the Higgs field is not in the adjoint representation, the scalar force usually being dominant.

A closely related subject is the interaction of superconducting vortices, which is the two dimensional analogue of the monopole problem, showing many similarities as well as some interesting differences. C.Rebbi describes some variational computations which show the importance of the limiting value $\lambda = 1$ of the parameter measuring the ratio of gauge and scalar coupling constants. For $\lambda < 1$ (type I region) vortices attract while for $\lambda > 1$ (type II region) they repel. For $\lambda = 1$ there appears to be no force between them. With regard to analytic results, in the $\lambda = 1$ limit a first order equation analogous to the Bogomolny one saturates the lower bound on energy which like monopoles is determined by the topological charge. Unlike monopoles, there exist rotationally symmetric solutions of arbitrary charge. The Atiyah-Singer theorem is used to calculate the number of parameters indexing the full set of multivortex solutions.

A.Trautman discusses gauge transformations as automorphisms of principal bundles which preserve the "absolute elements" of a field theory. In general relativity regarded as a gauge theory the only such absolute element is the canonical form on the linear frame bundle and therefore the gauge group in this sense consists of the canonical lift of the group of diffeomorphisms of the base space. In the presence of Higgs fields, regarded as sections of an associated bundle with fiber a homogeneous space, there is an associated reduction of the initial principal bundle on which the gauge fields are defined. The automorphism group of the reduced bundle is the group of symmetries of the Higgs fields.

Besides these papers summarizing most of the contents of the invited lectures there were several shorter contributed talks. Three of these by J.Isenberg, J.Tafel and L.Vinet have been included in detailed versions. The remaining contributed talks are described in abstracts. Unfortunately the text of G.Sparling's invited talk was not received in time to be included.

Although not all the material presented at the conference could be included, we hope that this volume communicates some of the richness of ideas that exist in the theory of gauge fields. The numerous unresolved problems which have been touched upon should provide a stimulus for further investigation in this fascinating area of research.

J.Harnad,
Centre de recherche de mathématiques appliquées,
Universite de Montréal.

S. Shnider,
Department of Mathematics,
McGill University

PHYSICAL EFFECTS OF INSTANTONS*

Claude Bernard
Department of Physics
University of California
Los Angeles, CA 90024, USA

1. Introduction

Instantons are Euclidean classical solutions of the Yang-Mills field equations--that is, they are non-trivial stationary points of the Euclidean Yang-Mills action. As such, they are the basis for a saddle point approximation of the Euclidean functional integral. The physical effects of instantons are just the contributions of these saddle points to the calculation of various physical processes.

Roughly speaking, there are two questions to be answered in deciding whether a particular saddle point gives an important contribution to a given integral: 1. Are there other saddle points whose contribution is numerically much larger? 2. How steeply does the integrand decrease in the neighborhood of the saddle point? A schematic representation of various possibilities is given in Fig. 1. We consider the integral $\int dx\, e^{-f(x)/g^2}$ where $f(x)$ has two minima: at $x = 0$, where $f(0) = 0$ (a "trivial" minimum), and at $x = x_o$ where $f(x_o) > 0$. For small g the situation is pictured in (a). The non-trivial minimum gives a negligible contribution compared to the trivial one. For large g (shown in (b)) the saddle point method breaks down entirely because the integrand is not peaked near 0 or x_o. Only in some intermediate situation like (c) (if it exists) can we say that the non-trivial minimum has important effects.

The question of importance of instantons in QCD is entirely analogous. There is, of course, the trivial minimum (all fields vanishing) and we must always compare instanton contributions to the contribution of this minimum (ordinary perturbation theory). The first step in such a comparison is the calculation of the ratio of the quadratic functional integral about the instanton to that about the vacuum. This calculation, originally performed by 't HOOFT for an SU(2) gauge theory [1], is described for an arbitrary SU(N) in Section 2. Some recent improvements in the calculation are included [2].

In Section 3 we briefly discuss the other limitation on the physical importance of instantons: how steeply the functional integrand falls near a particular saddle point. This is, of course, a difficult question to answer rigorously, but a minimal requirement for the validity of instanton calculations is provided by comparison with a particular "near-by" configuration: the cut-off meron pair. We describe the calculation by CALLAN, DASHEN and GROSS [3] of the point at which merons become important. This occurs before the instantons and anti-instantons become so closely packed that they lose their identity as individual saddle points (a breakdown of the "dilute gas limit").

Section 4 discusses the current status of certain calculations of the physical effects of instantons. In particular we describe work on instanton contributions to: e^+e^- annihilation at short distances [4-7], a confining phase transition in the presence of external color fields [8], and (briefly) quark mass generation [9,3]. We also make some remarks about the large N limit [10] and WITTEN's objections to instantons [11].

*Work supported in part by the National Science Foundation (US).

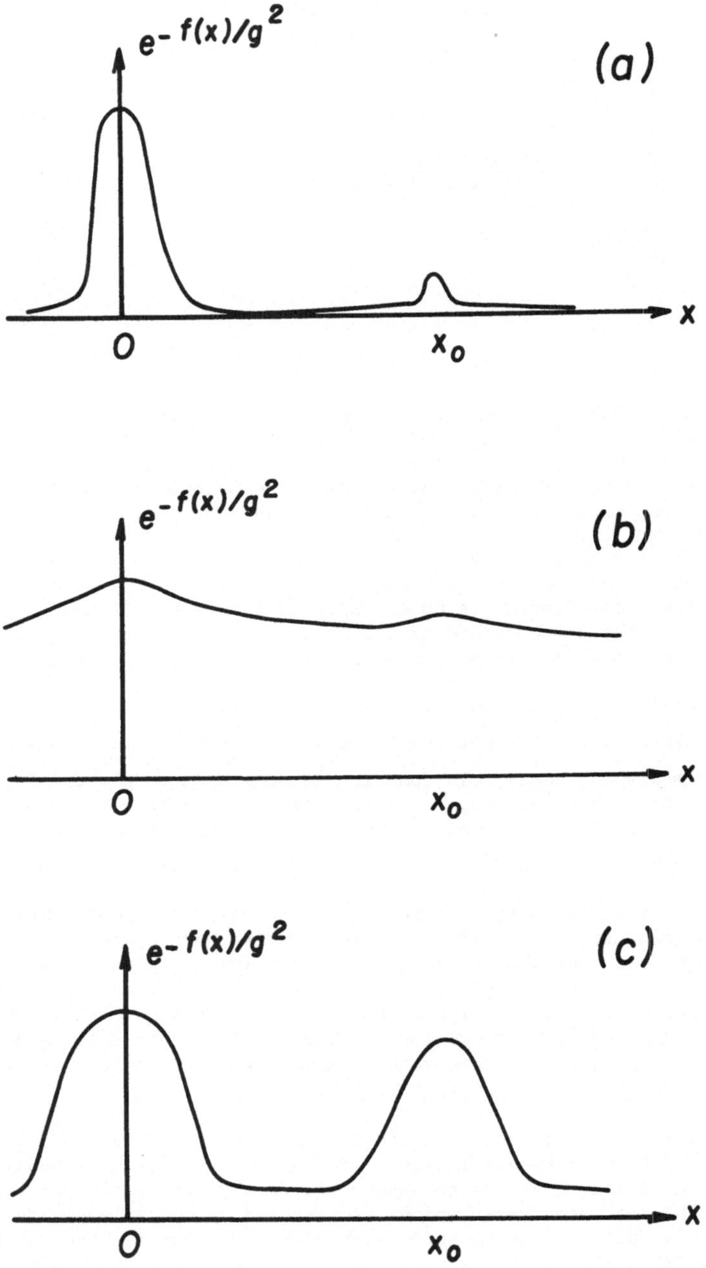

Fig. 1 A schematic representation of saddle point contributions to the integral $\int dx\, e^{-f(x)/g^2}$. The function f(x) is assumed to have two minima: a "trivial" one at x = 0 with f(0) = 0, and a "non-trivial" one at x_o with $f(x_o) > 0$.

(a) For small g the non-trivial minimum has a negligible contribution.

(b) For large g, the saddle point approximation breaks down.

(c) For intermediate g, the non-trivial minimum has important and calculable contributions.

It is important to emphasize that all calculations done to date on instanton effects in QCD are based on the view that superpositions of instantons and anti-instantons are the important (approximate) saddle points. Interactions among them may be included [8] (the gas of pseudoparticles may not necessarily be dilute in the strict sense); yet the configurations always have interparticle spacings which are greater than the particle size. Another possibility is suggested by the recent work of BERG and LUSCHER [12] and FATEEV, FROLOV, and SCHWARZ [13] on the O(3) σ-model in two dimensions. These authors find that it is very dense exact solutions of instantons (or anti-instantons) alone which give rise to the important physical effect--dynamical mass generation. (Superpositions of instantons and anti-instantons should, presumably, not be included in dense configurations since they are close to the vacuum.) Of course, the importance of such configurations has not been shown in QCD. We note, however, that the meron calculation discussed in Section 3 applies only to an isolated instanton and has nothing to say about the viability of dense configurations.

2. The Instanton Determinant

The first ingredient that goes into any instanton calculation is the value of $W^{(1)}$, the one-loop vacuum-vacuum amplitude about an instanton divided by the same amplitude about zero field. Expanding A_μ about its classical value

$$A_\mu = A_\mu^{cl} + A_\mu^{qu} , \tag{1}$$

we have, for the quadratic action

$$S = S^{cl} + \frac{1}{2} A^{qu} M_A A^{qu} + \phi^* M_{gh} \phi \tag{2}$$

where $S^{cl} = 8\pi^2/g^2$ and ϕ is the ghost field. $W^{(1)}$ is then given by

$$W^{(1)} = \int \prod_i d\gamma_i \; J(\gamma) \; Q(\gamma) \; e^{-8\pi^2/g^2} \tag{3}$$

where γ_i denote the collective coordinates of the instanton, $J(\gamma)$ is the collective coordinate Jacobian, and $Q(\gamma)$ is the ratio of determinants over non-zero modes:

$$Q(\gamma) \equiv \frac{\det^{-1/2} M_A(\gamma) \det M_{gh}(\gamma)}{[\det^{-1/2} M_A \det M_{gh}]_{A^{cl}=0}} . \tag{4}$$

't HOOFT [1] has calculated the determinants in $Q(\gamma)$ with Pauli-Villars regularization for fields of arbitrary spin and SU(2) isospin. Now, under the action of the SU(2) subgroup in which the instanton sits, the generators of SU(N) form one triplet, 2(N-2) doublets and N^2-4N+5 singlets [14]. The results of [1], which have been checked by others [15], then immediately give $Q(\gamma)$ for arbitrary SU(N) [2].

The evaluation of the collective coordinate Jacobian is somewhat more subtle. Let us first review the usual method [16] for introducing collective coordinates. For simplicity, consider a scalar field B which has a classical solution $B = B^{cl}(\gamma)$, where γ is a single collective coordinate. Let $M(\gamma)$ be the operator that appears in the expansion of the action to quadratic order about B^{cl}:

$$\begin{aligned} B &= B^{cl} + B^{qu} \\ S &= S^{cl} + \frac{1}{2} B^{qu} M B^{qu} \end{aligned} \tag{5}$$

M has a complete set of orthogonal eigenfunctions χ_i with eigenvalues ε_i and norms $\sqrt{u_i}$

$$u_i \equiv \langle \chi_i | \chi_i \rangle \ . \tag{6}$$

There is a zero mode:

$$\chi_o = \frac{\partial B^{cl}}{\partial \gamma} \ , \quad \varepsilon_o = 0 \ . \tag{7}$$

Expanding B^{qu},

$$B^{qu} = \sum_i \xi_i \chi_i \ , \tag{8}$$

the measure for functional integration is

$$(dB) = (dB^{qu}) = \prod_i \sqrt{u_i/2\pi}\, d\xi_i \ . \tag{9}$$

The overall normalization of (9) is arbitrary but conventional because it gives:

$$\int (dB)\ e^{-S} = \int \sqrt{u_o/2\pi}\, d\xi_o\ e^{-S^{cl}}\ \det{}^{-1/2} M + \ldots \tag{10}$$

where ... represents higher loops and other classical sectors. In other words, the factors of $1/\sqrt{2\pi}$ in (9) are chosen to cancel the factors of $\sqrt{2\pi}$ coming from Gaussian integrations so that we get precisely $\det^{-1/2}M$ and not $\det^{-1/2}M$ times some additional infinite factor. It was this normalization which was incorrect (or rather, inconsistent) in 't HOOFT's original paper [17].

We may now insert a factor of unity which will require the quantum field to be orthogonal to the zero mode:

$$1 = b \int d\gamma\ \delta\langle B - B^{cl}(\gamma) | \chi_o(\gamma) \rangle + \ldots \tag{11}$$

where ... represents terms of higher order in the quantum field and b is a constant determined by

$$b = \left\langle \frac{\partial B^{cl}(\gamma)}{\partial \gamma} \middle| \chi_o(\gamma) \right\rangle = u_o \ . \tag{12}$$

Inserting (11) into (10) and performing the integral over ξ_o gives

$$\int dB\ e^{-S} = \int d\gamma\ \sqrt{u_o/2\pi}\ e^{-S^{cl}}\ \det{}^{-1/2}M + \ldots \ . \tag{13}$$

The case of a gauge theory is similar but with one important difference: Because of the necessity of fixing a gauge, the derivatives of the classical field with respect to the collective coordinates will not, in general, be zero-modes, but will require additional gauge transformations to put them in the proper gauge. The ith zero mode is thus given by:

$$\psi_\mu^{(i)} = \frac{\partial A_\mu^{cl}}{\partial \gamma_i} + D_\mu^{cl} \Lambda^{(i)} \ . \tag{14}$$

This implies that the constant that appears in the step corresponding to (11) will not be the determinant of the matrix U,

$$U_{ij} \equiv \langle \psi^{(i)} | \psi^{(j)} \rangle \tag{15}$$

but rather, the determinant of V,

$$V_{ij} \equiv \left\langle \frac{\partial A^{cl}}{\partial \gamma_i} \,\middle|\, \psi^{(j)} \right\rangle . \tag{16}$$

This results in

$$J(\gamma) = \prod_i \frac{1}{\sqrt{2\pi}} (\det V)(\det U)^{-1/2} . \tag{17}$$

However if we choose to work in background gauge, so that

$$D_\mu^{cl} \psi_\mu^{(i)} = 0 , \tag{18}$$

and if the zero-modes and necessary gauge transformations fall off rapidly enough at large distances r,

$$\Lambda^{(i)} \psi_\mu^{(j)} < O\left(\frac{1}{r^3} \right) , \tag{19}$$

then a simple integration by parts gives V = U and the familiar result:

$$J(\gamma) = \prod_i \frac{1}{\sqrt{2\pi}} (\det U)^{1/2} . \tag{20}$$

't Hooft's original calculation was done in regular gauge where A_μ falls as 1/r. In this gauge the translation and dilation zero modes (which involve differentiation with respect to a dimensionful parameter) obey (19), but the gauge zero modes (which involve differentiation with respect to a dimensionless parameter) do not. Furthermore, even (17) cannot be used because the matrix V is divergent. This necessitates a very careful treatment of the gauge zero modes (placing the system in a box and changing the Fadeev-Popov ansatz). However, we can work in singular gauge A_μ falls as $1/r^3$ and all modes obey (19).

The advantage of singular gauge for treating the gauge zero modes can be easily seen. A gauge zero mode is a pure gauge transformation

$$\psi_\mu = D_\mu^{cl} \theta \tag{21}$$

which leaves A_μ in background gauge:

$$D_\mu^{cl} D_\mu^{cl} \theta = 0 . \tag{22}$$

In singular gauge, only the derivative terms in (22) matter at large distances so θ approaches a constant, which we may choose to be a particular group generator T^a. We can then write

$$\psi_\mu = D_\mu^{cl}[T^a + \Lambda] = [A_\mu^{cl}, T^a] + D_\mu^{cl} \Lambda \tag{23a}$$

or

$$\psi_\mu = \frac{\partial A_\mu^{c\ell}}{\partial t^a} + D_\mu^{c\ell}\Lambda \tag{23b}$$

where t^a is the parameter which describes infinitesimal global rotations of $A_\mu^{c\ell}$ in the direction T^a, and Λ is the remaining piece of θ which falls rapidly at large distances. This is precisely the form of (14), with (19) holding. In regular gauge, $A_\mu^{c\ell}$ competes with the derivatives in (22) so θ does not approach a constant. Therefore, Λ in (23) would not fall off with distance and (19) would be violated.

Thus in singular gauge we simply use (20) to calculate $J(\gamma)$. The zero modes are easily obtained. There are eight isospin 1 modes which are just the singular gauge version of the ones given in [1]; there are also 4(N-2) additional gauge zero modes corresponding to the generators which are in doublets under the action of the instanton's SU(2). The latter modes can be found from the isospin 1/2 spinor modes of [1], since vectors and right-handed spinors obey the same equation.

The results for $J(\gamma)$ and $Q(\gamma)$ can be put into (3). Because the integrand is independent of the gauge orientation of the instanton, the integral over those collective coordinates may be performed. This just gives the "number" of orientations of the instanton in SU(N) which is the volume of SU(N) divided by the volume of the stability subgroup which leaves the instanton invariant. The final result (with Pauli-Villars regularization) is [2]

$$W^{(1)} = \frac{4}{\pi^2} \frac{\exp[-\alpha(1)-2(N-2)\ \alpha(1/2)]}{(N-1)!\ (N-2)!} \int \frac{d^4z d\rho}{\rho^5} \left(\frac{4\pi^2}{g^2}\right)^{2N} e^{-8\pi^2/g^2(\rho)} \tag{24}$$

where z and ρ represent the instanton's location and scale, respectively, where the coefficients $\alpha(t)$ are tabulated in [1], and where, according to the renormalization group,

$$\frac{8\pi^2}{g^2(\rho)} = \frac{8\pi^2}{g^2} - \frac{11N}{3}\ \ell n(\mu_o\rho) \equiv -\frac{11N}{3}\ \ell n(\bar{\mu}\rho)\ . \tag{25}$$

Equation (24) differs from previous calculations [3,18]. For SU(3), it is a factor of 64 smaller (simply because we have taken cognizance of the recently discovered [17] error in 't HOOFT's calculations). For general SU(N), there is a more serious disagreement based on the identification of the stability group of the instanton. I will talk about the effect of changes in (24) on physical calculations later.

3. Breakdown of the Saddle Point Approximation

Superpositions of widely separated singular gauge instantons and anti-instantons are approximate saddle points of the functional integral. This "dilute gas limit" is the only limit in which instanton calculations have been done. If the dominant contribution to some physical process comes from the region where the instantons and anti-instantons are large in scale and closely packed, then we must say that the saddle point approximation has broken down, for such configurations are far from solutions. Higher order quantum corrections are then large and uncontrollable.

As a rough estimate to when such a breakdown occurs in QCD, one may first use (24) to find the mean density of SU(3) instantons of scale size ρ in the dilute gas approximation:

$$\frac{d\rho}{\rho^5} D(\rho) = b\,\frac{d\rho}{\rho^5}\, x^6\, e^{-x(\rho)}\ , \qquad b = .0015 \tag{26}$$

where $x = 8\pi^2/g^2$, $x(\rho) = 8\pi^2/g^2(\rho)$. One can then calculate $x(\rho_D)$ where ρ_D is the scale at which instantons of size $\rho \leq \rho_D$ occupy all of space-time. If we assume, in the usual way [3], that higher order corrections change x^6 in (2.6) to $(x(\rho))^6$ (there is no real justification for this--the scale at which x is evaluated is presumably process dependent) and then blindly use (25), we find $x(\rho_D) \simeq 0$. Of course, one is hardly justified in using (25) down to such values of $x(\rho)$--the point is merely that this limit on the saddle point approximation is no stricter than the requirement that $g(\rho)$ be small enough for ordinary perturbation theory. This is in contrast with the result using the old value [3] ($b = .1$); in that case $x(\rho_D) \simeq 14$, where one might expect perturbation theory still to be good.

A more stringent condition on the validity of the saddle point approximation is obtained by examining meron configurations. CALLAN, DASHEN, and GROSS [3] show that the one-loop action of a pair of cut-off merons, separated by a distance d, goes like

$$S^{(1)} = \left(\frac{3}{4}\, x(d) - 6.55 \right) \ell n(\bar{\mu} d) + \ldots \tag{27}$$

where ... represents terms that do not depend on d. They argue that (27) implies a phase transition, with a gas of free merons present when

$$\frac{\partial S^{(1)}}{\partial(\ell n d)} < 8 \, . \tag{28}$$

One can find (28) as the condition that the susceptibility in an external color field diverges; a less-rigorous, but quicker derivation just comes from considering a single meron. A single meron has divergent action, so we put it in a box of radius R. The one-loop action is then presumably half of (27) with d replaced by R [19]. The entropy of position of this meron is proportional to R^4. The condition (28) is then just the requirement that the free energy of a meron become negative in the infinite volume limit.

Equation (28) provides a value of the coupling constant for which instantons become unstable for break-up into merons. In other words, instanton calculations are unreliable when they involve scales greater than ρ_M, where

$$x(\rho_M) \simeq 17 \, . \tag{29}$$

Three further remarks are in order:

1. The coupling constant in (29) is defined with Pauli-Villars regularization; with another regularization scheme and/or definition of the coupling constant, $x(\rho_M)$ will of course have a different numerical value.

2. Unlike $x(\rho_D)$, $x(\rho_M)$ is totally independent of the numerical value of the constant b within any one regularization scheme. It just depends on how the one-loop determinant around two merons changes with the distance, not on any numerical constant in front of the instanton or meron result.

3. Even if one does not believe that a gas of free merons can exist (no one has written down such a classical configuration; furthermore, the calculation of [3] develops problems--negative eigenvalues of the quadratic operator--for large meron separation), (29) should still be a reasonably reliable indication of where the saddle point approximation breaks down, to be superceded by other non-perturbative techniques.

4. Physical Effects

We start by examining some calculations of instanton effects on the short distance behavior of hadronic current correlation functions [4,5,6]. In particular, we may consider $\Pi(p^2)$, defined in terms of two electromagnetic currents:

$$(p^2 g_{\mu\nu} - p_\mu p_\nu)\ \Pi(p^2) = \int d^4x\ e^{ipx} \langle 0|T(J_\mu(x)\ J_\nu(0)|0\rangle \tag{30}$$

in order to get information about the total e^+e^- annihilation cross section into hadrons. The calculation of the leading instanton effects on $\Pi(p^2)$ is fairly straightforward once one has the massless Fermion propagator in the presence of a single instanton field [20]. The massive propagator can then be obtained order by order in the mass. The result for $\Pi(p^2)$ [4,5] has large contributions from instantons of scales greater than ρ_M, indicating a breakdown of the saddle point approximation. On the other hand, if we simply cut off the integral over instanton sizes at ρ_M, then the effects of instantons are much smaller than the first and second order perturbation theory corrections to Π [21]. Furthermore, we could calculate a quantity for which the saddle point approximation is truly reliable (with only small instantons of size $\rho \sim 1/p$ contributing)--either the imaginary part of the naive continuation of $\Pi(p^2)$ to timelike p [2,5], or the Fourier transform of $x^2\Pi(x)$ [6]. In both cases, the instanton contribution is truly tiny compared to perturbative contributions for p^2 large enough so that the whole scheme makes sense. Of course, for small p^2 instanton effects become large; but this is where the approximation is breaking down. We can only say that something (i.e. confinement) is going on at small p^2; we cannot calculate anything reliable about that region.

A more exciting possibility is advocated in the recent work of CALLAN, DASHEN, and GROSS [8] on the role of instantons in quark confinement and the formation of a hadron bag. The starting point for this work is the observation that for an instanton in singular gauge, the commutator term in $F_{\mu\nu}$ falls off faster than the derivative terms, so that a sufficiently dilute gas of instantons interacts in an essentially Abelian way. We can replace $F_{\mu\nu}$ by $\mathcal{F}_{\mu\nu}$ defined as

$$\mathcal{F}_{\mu\nu} = \partial_\mu A_\nu - \partial_\nu A_\mu\ . \tag{31}$$

The instanton can be thought of as creating Abelian field through the equation

$$\partial_\mu \mathcal{F}_{\mu\nu} = j_\nu\ , \tag{32}$$

where j_ν represents all the other terms in the non-Abelian equation of motion. We can designate the instantons as "permanent magnetic dipoles" in four dimensions: "Permanent" because j_ν in (32) exists in the absence of external sources; "magnetic" because we have a static situation in five dimensions, and $\mathcal{F}_{\mu\nu}$ consists of the "space-space" components of the curl of A; "dipole" because $\mathcal{F}_{\mu\nu}$ falls as $1/r^4$, and a monopole field falls as $1/r^3$ in four dimensions.

The gas of instantons is then a "paramagnetic" medium of dipoles, which will line up with an external color field. The permeability μ, defined by,

$$\mathcal{F}_{\mu\nu} = \mu\ \mathcal{F}_{\mu\nu}^{ext} \tag{33}$$

will be greater than one. (Equation (33) is conveniently written as

$$E = \mu D \tag{34}$$

where E is an i-4 component of $F_{\mu\nu}$ and D is the corresponding component of $F_{\mu\nu}^{ext}$.)

Callan et al. then argue that the instanton medium creates a D vs. E phase diagram which looks like Fig. 2. The instability in this curve at point 1, where $\partial D/\partial E$ changes sign, is taken as an indication that there is a phase transition, as E decreases, from a phase with μ small to a phase, as yet inaccessible to calculation, with μ large or infinite. This would describe the formation of a hadron bag which confines quarks--at the edge of the bag, the external field created by the quarks would be at the critical value; outside the bag would be a medium of infinite μ which expells the field entirely.

Before commenting on the validity of the saddle point approximation in this application, I will briefly sketch the physical reasons for the instability that appears in Fig. 2. The argument will be very rough and qualitative. In the presence of an external field D, the free energy of the gas is changed from F_o to F, where [22]

$$F(n,D) = F_o + \frac{1}{2} DE = F_o + \frac{1}{2} \mu D^2 . \tag{35}$$

Since μ-1 is roughly proportional to n, the instanton density, (35) shows that the presence of D will lower the density: minimizing F must be done by lowering μ (and hence n) since F_o is stationary to first order. This is the phenomenon of "magnetostriction." Doing the thermodynamics, we find that the functional form is, roughly,

$$n \sim n_o e^{-D^2} \tag{36}$$

where n_o is the density in the absence of D. The proportionality of μ-1 and n then gives

$$E = \mu D \sim (1 + ce^{-D^2}) \cdot D , \tag{37}$$

where c is a proportionality constant. For c large enough this function has the shape of Fig. 2. (As D decreases from large values, μ increases so rapidly that E actually increases.)

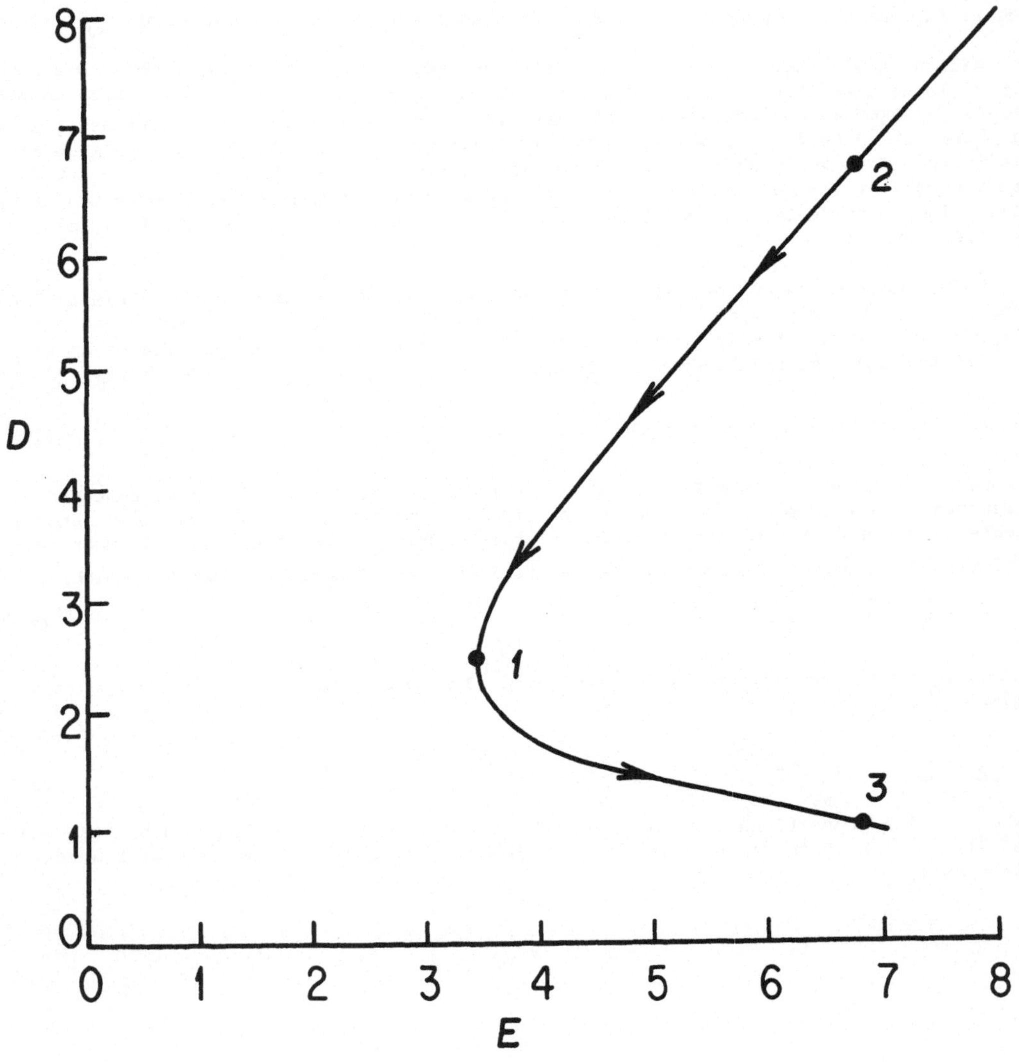

Fig. 2 D vs. E (measured in units of $\bar{\mu}^{2}$, the square of the renormalization mass). The arrows represent the direction of decreasing x (increasing coupling constant). Points 1, 2, and 3 are referred to in the text. (The calculation takes into account instanton interactions by placing each instanton in a spherical cavity of radius $R = 2.2\rho_{peak}$ in the permeable medium formed by other instantons. As in Ref. 8, other choices for R do not change the result significantly.)

The calculation in [8] has the virtue that the presence of the external field cuts off the integral over instanton sizes. Large instantons, which have large dipole moments, are suppressed. As D is decreased, the scale of instantons which contribute the most to μ increases, and hence $x(\rho)$ decreases. Unfortunately the scale at which instantons ionize into merons ($x(\rho_M) \simeq 17$) occurs at the point 2 on Fig. 2, where $\mu \simeq 1.01$ and nothing much of interest is happening. At the instability, point 1, $x(\rho) \simeq 12$, far below where the saddle point approximation may be trusted. To carry out a similar calculation for a gas of free merons seems very difficult. Even were such a configuration available, an Abelian calculation would presumably be impossible: the slow fall-off of the meron field makes the commutator in $F_{\mu\nu}$ competitive with the derivative terms. We note that in the original version of this work, where the old value of b = .1 in (26) was used, meron ionization occurred at point 3 on Fig. 2, which would have allowed the calculation of the instability to be reliable. At present, one may take it only as a qualitative indication of what the non-perturbative sector may do.

Very recently, the same authors have argued that instantons produce the transition between the region of weak coupling (ordinary perturbation theory) and the region of strong coupling (carried out on a lattice) [23]. They find that renormalization effects due to instantons become important in the range of coupling constant $g \simeq 1.5$-3 and may cause a sharp transition between the two regions. However, when I translate $x(\rho_M) = 17$ to the lattice definition of the coupling constant used in their work, I find $g(\rho_M) \simeq 1.5$. This indicates that the meron problem with the bag calculation occurs here, too.

One other application of instantons which I briefly mention is the breaking of chiral SU(N) × SU(N) and the dynamical generation of quark masses [9]. This would take place in the usual NAMBU--JONA-LASINIO [24] fashion, through the self-consistent action of the non-local, 2N Fermion interaction term [1] induced by the Fermion zero mode about the instanton.

Such calculations involve instantons of all scales, and cannot be expected to produce reliable, quantitative results for dynamically generated quark masses. However, CARLITZ and CREAMER [25] have recently argued that the form of generated interaction has certain predictive power. In particular, they find differences in transverse momentum distributions of the up and down quarks within a proton. Such possibilities deserve further study.

Finally, I would like to comment on WITTEN's objections to instantons [11]. These stem mainly from consideration of the large N, planar limit [10] in which the number of colors, N, goes to infinity with g^2N fixed. In this limit, dilute gas instanton effects vanish [26,11] as e^{-N} as is easily seen from (24) [27]. Thus, if one believes that all important physics can be seen at large N, then one must, perforce, believe that dilute gas instantons are not important. However, it seems to me that one may appreciate the successes of large N limit (Regge phenomenology, Zweig's rule etc.) without insisting that all physics is present at large N--just as one may appreciate the successes of perturbation theory in g without insisting on the absence of non-perturbative (e^{-1/g^2}) effects.

A further objection to instantons in this context comes from the mass of the "η"--the missing isosinglet Goldstone boson. This must come from non-vanishing zero momentum matrix elements of the operator $F_{\mu\nu}\tilde{F}_{\mu\nu}$, which appears on the "right-hand side" of the anomaly. Dilute gas instantons, of course, contribute to these matrix elements, making the mass of the η go like e^{-N}. But this is in disagreement with the naive quark model, where the η mass comes from the annihilation of quark-antiquark into gluons and is of order 1/N. Witten shows that agreement with the quark model may be obtained if there are contributions to zero momentum matrix elements of $F_{\mu\nu}\tilde{F}_{\mu\nu}$ coming from the sum of all orders of perturbation theory. (Since $F_{\mu\nu}\tilde{F}_{\mu\nu}$ is a total divergence this cannot come in any finite order.) Such a scenario is not impossible because it happens in a two-dimensional model; still, in QCD, it is only a guess. Compared to the dilute gas instanton picture, it does have the advantage of reconciling (in orders in N) two calculations of the η mass.

However, recent work by JEVICKI [28] has shown how misleading it is to say that instantons disappear in the large N limit. In the two-dimensional CP^{N-1} σ models, Witten finds the large N behavior by first inserting a Lagrange multiplier field λ to enforce the non-linear constraint, then performing the functional integral over the scalar fields, and finally performing the integral over λ with a saddle point approximation. Jevicki points out that if, instead, one deforms the λ contour to pick up the contribution of poles of the integrand by a "functional residue theorem," then the residues give precisely the instanton contributions. Since dense pseudoparticle configurations may be involved, there is no contradiction with the above remarks on the dilute gas at large N. This is in fact what happens in CP^1 (the O(3) σ model) [12, 13]. Thus instantons are not antithetical to the large N limit. Witten's work, taken in conjunction with the general failure to find important physical effects of the instanton gas, does make me rather pessimistic about the future of such instanton gas calculations. However, there is still hope that <u>dense</u> configurations will be significant in QCD.

References

1. G. 't Hooft, Phys. Rev. D14 (1976) 3432.
2. C. Bernard, Phys. Rev. D19 (1979) 3013.
3. C. Callan, R. Dashen, and D. Gross, Phys. Rev. D17 (1978) 2717.
4. N. Andrei and D. Gross, Phys. Rev. D18 (1978) 468.
5. L. Baulieu, J. Ellis, M. K. Gaillard, and W. J. Zakrezewski, Phys. Lett. 77B (1978) 290.
6. T. Appelquist and R. Shankar, Phys. Rev. D18 (1978) 2952.
7. Other treatments, which we do not discuss here, include R. Carlitz and C. Lee, Phys. Rev. D17 (1978) 3238 and M. Suzuki, Phys. Lett. 76B (1978) 466.
8. C. Callan, R. Dashen, and D. Gross, Phys. Rev. D19 (1979) 1826.
9. C. Callan, R. Dashen, and D. Gross, Phys. Rev. D16 (1977) 2526; D. Caldi, Phys. Rev. Lett. 39 (1977) 121; R. Carlitz, Phys. Rev. D17 (1978) 3225.
10. G. 't Hooft, Nucl. Phys. B72 (1974) 461.
11. E. Witten, "Instantons, the Quark Model, and the 1/N Expansion," Harvard preprint HUTP-78/A042 (1978) and "Current Algebra Theorems for the U(1) Goldstone Boson," HUTP-79/A014 (1979).
12. B. Berg and M. Lüscher, "Computations of Quantum Fluctuations Around Multi-Instanton Fields from Exact Green's Functions: the CP^{n-1} Case," DESY preprint, March 1979.
13. V. A. Fateev, I. V. Frolov, and A. S. Schwarz, ITEP preprint, 1979.
14. C. Bernard, N. Christ, A. Guth, and E. Weinberg, Phys. Rev. D15 (1977) 2967.
15. F. R. Ore, Jr., Phys. Rev. D16 (1977) 2577; S. Chadha, A. D'Adda, P. DiVecchia, and F. Nicodemi, Phys. Lett. 72B (1977) 103.

16. J. Gervais and B. Sakita, Phys. Rev. D11 (1975) 2943; E. Tomboulis, ibid. 12 (1978) 1678.
17. G. 't Hooft, Phys. Rev. D18 (1978) 2199.
18. Yu. Bashilov and S. Pakrovski, Nuc. Phys. B143 (1978) 431.
19. This can be proved in the classical case, at least, by a conformal transformation. See [3].
20. L. Brown, R. Carlitz, D. Creamer, and C. Lee, Phys. Rev. D17 (1978) 1583.
21. The results of [4] have been adjusted for the change in the parameter b.
22. Although the sign of the added term in (35) appears to be the conventional one, it is produced, in fact, from the cancellation of two changes of sign with subtle origins: (a) There is a minus sign coming from the fact that we are really doing magnetostatics in four dimensions. This means we should start with the thermodynamic potentials involving B and H in which the role of external field and total field are interchanged. (b) There is a minus sign coming from the fact that D is the external electric field produced by quark charges in Minkowski space, which becomes imaginary in Euclidean space.
23. C. Callan, R. Dashen, D. Gross, "Instantons as a Bridge between Weak and Strong Coupling in QCD," Princeton preprint, August 1979.
24. Y. Nambu and G. Jona-Lasinio, Phys. Rev. 122 (1961) 345.
25. R. Carlitz and D. Creamer, "Instanton Induced Interactions," Pittsburgh preprint PITT-208 (1979).
26. J. Koplik, A. Neveu, and S. Nussinov, Nuc. Phys. B123 (1977) 109.
27. Differences between (24) and [26] and [18] do not change this conclusion.
28. A. Jevicki, "Collective Behavior of Instantons in QCD," Institute for Advanced Study preprint, June 1979.

SOME ASPECTS OF INSTANTONS*

E. Corrigan
California Institute of Technology, Pasadena, California 91125
and
Department of Mathematics, University of Durham, Durham, U.K.**
and
P. Goddard
Department of Applied Mathematics and Theoretical Physics,
University of Cambridge, Cambridge, U.K.

Contents

* Work supported in part by the Alfred P. Sloan Foundation
** present address

1. Instantons and zeta functions

1.1 Instantons as local minima of a Euclidean action

In these lectures we should like to discuss, fairly sketchily, two types of theories Yang-Mills gauge theory [1] in four dimensions and certain σ-models [2], the so-called CP^{n-1} models, in two dimensions. It will be difficult to go into great detail in the short time available but our intention is to give some inkling of the ideas and sorts of calculations involved in the estimation of the functional integrals presumed to define each type of theory. Inevitably almost, the two dimensional calculation is at the moment more explicit and instructive. However, there is really no reason to believe that the answer to that problem has anything to do with the answer to the four dimensional problem - only time and hard work can decide that.

The typical Euclidean function integral [3] in which we are interested is of the form,

$$Z = \frac{1}{Z_o} \int d[\phi]\; e^{-\frac{1}{g^2} S_\phi} \Phi(\phi) \tag{1.1}$$

and generates by a suitable choice of Φ all the Euclidean Green functions of the theory. For small values of g the integral can be replaced, to leading order in g, by a sum of gaussian integrals centred at the minima of the action S_ϕ. More correctly, the integral in (1.1) has an asymptotic expansion as $g \to 0$ and our efforts are directed towards picking out the leading term.

To see what the leading term looks like consider first a finite dimensional integral [4]

$$I = \int d^n x \quad f(x) \quad e^{-\frac{1}{g^2} S(x)} \quad . \tag{1.2}$$

Suppose that the minimum of $S(x)$ occurs on a k dimensional set of points M, parameterised by $x(t_1 \ldots t_k)$, say, such that

$$S(M) = S_o$$

$$\left.\frac{\partial S}{\partial x_i}\right|_M = 0, \quad \left.\frac{\partial^2 S}{\partial x_i \partial x_j}\right|_M = C_{ij} \; (t_1 \ldots t_k). \tag{1.3}$$

Then as $g \to 0$ the leading contribution to I is written as an integral over M,

$$I \sim g^{n-k} \; (2\pi)^{\frac{n-k}{2}} \int \prod_1^k dt_i \; \sqrt{\mathfrak{N}} \quad f_o \; e^{-\frac{1}{g^2} S_o} \; (\det{}'C)^{-\frac{1}{2}} \tag{1.4}$$

where

$$\mathfrak{N} = \det\left(\sum_i \frac{\partial x_i}{\partial t_\ell} \frac{\partial x_i}{\partial t_m} \right) \tag{1.5}$$

and f_o denotes $f(x)$ restricted to M and is, therefore, a function of the t's. Det'C is the product of the non-zero eigenvalues of the matrix C_{ij}. It is clear from the derivation of eq.(1.4) that contributions to the next order in g will be very much more complicated to compute.

The field theory approximation is similar to eq.(1.4) except that the measure has to be normalised correctly so that (i) no factors of $(\sqrt{2\pi}g)^n$, $n \to \infty$, occur and (ii) we only have to compute determinants of dimensionless quantities. To achieve (ii) a parameter μ of dimensions inverse length has to be introduced; this is not crucial for finite dimensional integrals where overall factors can be distributed at will in the integrand of eq.(1.4). Thus, a more appropriate finite dimensional analogy is actually

$$\int d^n x \left(\frac{\mu^2}{2\pi g^2} \right)^{n/2} f(x) e^{-\frac{1}{g^2} S} \sim \left(\frac{\mu}{g\sqrt{2\pi}} \right)^k e^{-\frac{1}{g^2} S_o} \int \prod_1^k dt_i \; \sqrt{\mathfrak{N}} \, f_o (\det{}'[C/\mu^2])^{-\frac{1}{2}}. \tag{1.6}$$

This can be taken over to the field theories in which we are interested since it turns out that the set of parameters describing the minima of the action is finite dimensional.

For example, a gauge theory with gauge group G is described by a vector potential A_μ and a field strength $F_{\mu\nu}$ [1],

$$F_{\mu\nu} = \partial_\mu A_\nu - \partial_\nu A_\mu + [A_\mu, A_\nu] \; , \tag{1.7}$$

both taking values in the Lie algebra of G and transforming under elements of G as,

$$A_\mu \to g(x)^{-1} A_\mu \, g(x) + g(x)^{-1} \partial_\mu \, g(x) \tag{1.8}$$

$$F_{\mu\nu} \to g(x)^{-1} F_{\mu\nu} \, g(x) \tag{1.9}$$

(A_μ and $F_{\mu\nu}$ are antihermitian). An appropriate gauge invariant Euclidean action, S, is given, in the usual way, by

$$S = -\tfrac{1}{2} \int d^4 x \; \mathrm{Tr}(F_{\mu\nu} F_{\mu\nu}). \tag{1.10}$$

It was pointed out by BELAVIN, POLYAKOV, SCHWARZ and TYUPKIN [5] that if we restrict ourselves to a consideration of only those vector potentials which are pure gauges at Euclidean infinity (i.e. $F_{\mu\nu} \to 0$ there, which is necessary for S to be finite) then vector potentials fall into equivalence classes corresponding to the third homotopy group of G, $\pi_3(G)$. These equivalence classes are conveniently labelled by integers k which can be computed analytically using

$$8\pi^2 k = -\tfrac{1}{2} \int d^4x \, \mathrm{Tr}(F_{\mu\nu} \, {}^*F_{\mu\nu}) \qquad k=0, \pm 1, \pm 2 \ldots \tag{1.11}$$

where

$$^*F_{\mu\nu} = \tfrac{1}{2} \varepsilon_{\mu\nu\rho\sigma} F_{\rho\sigma}. \tag{1.12}$$

Furthermore, since

$$S = -\tfrac{1}{4} \int d^4x \, \mathrm{Tr}[(F \pm F^*)^2 \pm 2F_{\mu\nu} \, {}^*F_{\mu\nu}]$$

the action is clearly bounded below in each equivalence class by $8\pi^2|k|$ and the bound is reached when

$$F_{\mu\nu} = \pm {}^*F_{\mu\nu} \tag{1.13}$$

i.e., when the field strength $F_{\mu\nu}$ is self-dual or anti-self-dual, respectively.

The problem of finding all vector potentials satisfying eq.(1.13) for a given integer k and gauge group G has drawn interest from both physicists and mathematicians and an elegant solution due to ATIYAH, DRINFELD, HITCHIN and MANIN [6,7] will be described and discussed in section 2. Certainly, we shall need to know a great deal about these solutions (instantons) if we wish to study the functional integral and estimate it via eq.(1.6).

The gauge theory analogue of the argument leading to eq.(1.4) in the finite dimensional case runs as follows [4]. An arbitrary potential A_μ is conveniently split into three pieces

$$A_\mu = A^o_\mu + D^o_\mu \phi + a_\mu \tag{1.14}$$

where A^o_μ is an instanton potential depending upon a number, N(k), of parameters t_i, D^o_μ is the covariant derivative constructed from A^o_μ, viz,

$$D^o_\mu \phi = \partial_\mu \phi + [A^o_\mu, \phi] \tag{1.15}$$

and a_μ, $D^o_\mu \phi$, $\frac{\partial A^o_\mu}{\partial t_i}$ $i = 1 \ldots . N(k)$ are chosen orthogonal to each other, i.e.

$$\int a_\mu D^o_\mu \phi \, d^4x = 0 \tag{1.16}$$

(The functional integral over ϕ will give an infinite factor to be absorbed into Z_o. It is just the 'volume' of the infinite dimensional group of all gauge transformations). The expansion of the action up to terms quadratic in a_μ is

$$S = 8\pi^2|k| + \int d^4x \, \mathrm{Tr}(a_\mu \Delta_{\mu\nu} a_\nu) + O(a^3) \tag{1.17}$$

where

$$\Delta_{\mu\nu} a_\nu = (D^o_\nu)^2 a_\mu + 2[F^o_{\mu\nu}, a_\nu] - D^o_\nu D^o_\mu a_\nu. \tag{1.18}$$

The Jacobian (the factor $\sqrt{\pi}$ in eq.(1.4)) here has two pieces one from the finite dimensional set of parameters t_i,

$$[\det \{ \int d^4x \, \mathrm{Tr} \frac{\partial A^o_\mu}{\partial t_i} \frac{\partial A^o_\mu}{\partial t_j} \}]^{\frac{1}{2}}, \tag{1.19}$$

and one from the functional integral over ϕ,

$$[\det(-[D^o]^2/\mu^2)]^{\frac{1}{2}} \qquad (1.20)$$

Finally, the analogue of $\det'[{}^C/\mu^2]^{-\frac{1}{2}}$ is

$$[\det'(-\Delta_{\mu\nu}/\mu^2)]^{-\frac{1}{2}} \qquad (1.21)$$

and again the prime denotes that only the non-zero eigenvalues are to be counted in the computation of the determinant.

To understand a little more about the relationship between the operators D^2 and $\Delta_{\mu\nu}$ consider the sets of fields V_0, V_1, V_2 scalars, vectors, anti-self-dual tensors (we are thinking of $k > 0$ now) respectively in the adjoint representation of the gauge group G. (The vectors are to be thought of not as gauge fields like A^o_μ but as transforming homogeneously under G like the pieces $D^o_\mu\phi$ and a_μ of eq.(1.14)). Consider also the sequence of differential operators relating the sets of fields [8,4],

$$V_0 \underset{d_1}{\rightarrow} V_1 \underset{d_2}{\rightarrow} V_2$$

defined by

$$(d_1V)_\mu = D^o_\mu V$$

$$(d_2V)_{\mu\nu} = \tfrac{1}{2}(\delta_{\mu\rho}\delta_{\nu\sigma} - \delta_{\nu\rho}\delta_{\mu\sigma} - \varepsilon_{\mu\nu\rho\sigma})\, D^o_\rho V_\sigma . \qquad (1.22)$$

Then

$$F^o_{\mu\nu} = {*F^o_{\mu\nu}} \quad \text{if and only if} \quad d_2d_1 = 0$$

and

$$(d_2^* \, d_2 \, V)_\mu = -\Delta_{\mu\nu}V_\nu \qquad (1.23)$$

where the * in this and subsequent equations denotes the adjoint. There is a 'Laplacian' appropriate to each space V_i :

$$d_1^*d_1 \equiv \Delta_0, \quad d_1d_1^* + d_2^*d_2 \equiv \Delta_1, \quad d_2d_2^* \equiv \Delta_2. \qquad (1.24)$$

It is not difficult to show that $d_2{}^*d_2$ and $d_2d_2{}^*$ have the same <u>non-zero</u> eigenvalues, occurring with the same multiplicity. So the ratio of determinants appearing in the integrand is just

$$(\det'[\Delta_o/\mu^2]\,/\,\det'[\Delta_2/\mu^2])^{\frac{1}{2}}, \qquad (1.25)$$

if we put together the factors (1.20), (1.21) and the above remark. Furthermore, remembering that $d_2d_1 = 0$, it is easy to convince one's self that

$$\det'[\Delta_1/\mu^2] = \det[\Delta_o/\mu^2]\,\det[\Delta_2/\mu^2] \qquad (1.26)$$

and the ratio (1.25) becomes

$$\det'[\Delta_o/\mu^2]\,/\,\det'[\Delta_1/\mu^2]^{\frac{1}{2}},$$

where

$$(\Delta_1)_{\mu\nu}\, a_\nu = -[D_o]^2 a_\mu - 2[F_{\mu\nu}, a_\nu] , \qquad (1.27)$$

which is the expression which usually occurs in the physics literature, incorporating the FADEEV-POPOV [9] factor in the 'background' gauge $D^o_\mu a_\mu = 0$. However, as first pointed out by AMATI and ROUET [10] and emphasised by others [11] we have not picked a unique global gauge in which to perform the functional integral.

In Euclidean space we may use that the fact that $F^{o}_{\mu\nu}$ is self-dual to show that Δ_z is just $-(D^o)^2$, the same as Δ_o, but acting on a space three times the size (anti-self-dual tensor rather than scalars). So

$$\det{}' [\Delta_2/\mu^2] = (\det{}' [\Delta_o/\mu^2])^3$$

and (1.28)

$$\det{}' [\Delta_1/\mu^2] = (\det{}' [\Delta_o/\mu^2])^4,$$

the latter equality following from the former and eq.(1.26).

Putting all the pieces together we have in general to evaluate

$$\left(\frac{\mu}{\sqrt{2\pi}\, g} \right)^{N(k)} e^{-\frac{8\pi^2 k}{g^2}} \int \prod_{1}^{N(k)} dt_i \; \frac{\det{}'(-(D^o)^2/\mu^2)}{(\det{}'(\Delta_1/\mu^2))^{\frac{1}{2}}} \sqrt{\mathfrak{N}'}\; \Phi \tag{1.29}$$

with

$$\mathfrak{N}' = \det \left(\int d^4x \; \frac{\partial A^o_\mu}{\partial t_i} \frac{\partial A^o_\mu}{\partial t_j} \right)$$

to obtain for each k the leading contribution to the asymptotic expansion as $g \to 0$. To do so we need to learn how to compute the functional determinants.

1.2 Determinants and zeta functions

There have been many ways of defining determinants developed by field theorists but essentially two have been used in instanton calculations, a PAULI-VILLARS type scheme [12,13,14,31] and the zeta function (or proper-time) method [4,15-19,20]. The latter seems to us to have a number of advantages particularly for discussing conformal properties of the determinants [4,21,22], although it has not yet been shown to be part of a consistent scheme for defining the Green functions of the theory to all orders in the coupling constant. As HAWKING has emphasised [19], the zeta function method has the particular advantage of being equally useful in discussions of curved spaces and indeed the mathematical development of the notion of determinant was chiefly concerned with differential operators defined on Riemannian manifolds [17]. Since the technique is not all that well known it is perhaps worth digressing to discuss it.

Suppose the finite N x N matrix A is hermitian and positive definite with eigenvalues $\lambda_1 \dots \lambda_N$ (not necessarily distinct). We may write

$$\zeta_A(s) = \sum_{1}^{N} \lambda_n^{-s}, \tag{1.30}$$

which defines a function analytic in s with the two important (and obvious) properties

(i) $\zeta_A(0) = N$, the dimension of A, (1.31a)

(ii) $\zeta'_A(0) = -\ln \det A$. (1.31b)

Simple though these statements are, they form the basis of a <u>definition</u> of functional determinants and the dimension of operators appearing, for example, in eq.(1.29).

For a differential operator such as $-(D^o)^2$ (which is positive definite if we work on the sphere S^4 which is conformally related to the flat Euclidean space R^4) with an infinite set of eigenvalues $\lambda_1, \lambda_2, \dots$. We can also define

$$\zeta_{-D^2}(s) = \sum_{1}^{\infty} \lambda_n^{-s} \tag{1.32}$$

and immediately run into trouble with $\zeta_{-D^2}(0)$. However, the series in eq.(1.32) is only defined for Re $s > 2$ (typically). To analytically continue down to $s = 0$ we borrow a technique from the analysis of the Riemannian zeta function [23] where we would write

$$\zeta(s) = \sum_1^\infty \frac{1}{n^s} = \frac{1}{\Gamma(s)}\int_0^\infty dt\; t^{s-1}\; \frac{1}{e^t-1}\,. \tag{1.33}$$

The integral on the right hand side of eq.(1.33) is suitable for evaluating the analytic continuation of ζ to all complex s and reveals a pole at s = 1. By expanding the integrand in eq.(1.33) near t = 0, one easily obtains $\zeta(0) = -\frac{1}{2}$; with more effort one obtains $\zeta'(0) = -\frac{1}{2}\ln 2\pi$. (Actually, the latter can be checked straightforwardly by evaluating the path integral identity

$$\frac{1}{\sqrt{4\pi}\, t}\, e^{-\frac{|x-y|^2}{4t}} = \int d[x]\;\; \exp\left(-\tfrac{1}{4}\int_0^t \dot{x}^2 d\sigma\right) \tag{1.34}$$

as a determinant involving the Riemann zeta function, via eq.(1.31b)).

The analogue of eq.(1.33) for the finite matrix A is just,

$$\zeta_A(s) = \frac{1}{\Gamma(s)}\int_0^\infty dt\; t^{s-1}\; \mathrm{Tr}(e^{-At}) \tag{1.35}$$

and, in turn, this is generalisable to differential operators such as $-(D^o)^2$ by noting that the analogue of e^{-At} is " $e^{(D^o)^2 t}$ " or, more correctly, $\mathcal{G}(x,y;t)$ defined to be the solution to

$$\frac{\partial}{\partial t}\mathcal{G}(x,y;t) = (\overrightarrow{D}^o)^2\; \mathcal{G}(x,y;t) = \mathcal{G}(x,y;t)\; (\overleftarrow{D}^o)^2 \tag{1.36}$$

with the boundary condition

$$\mathcal{G}(x,y;t) = \delta(x-y) + O(t) \qquad \text{as} \qquad t \to 0 \tag{1.37}$$

(Note: the arrow above an operator indicates which variable x or y it acts upon; thus $\overrightarrow{D}^o \equiv \partial_\mu + [A_\mu,]$, etc.). Notice that in the gauge theory the function $\mathcal{G}$ transforms as

$$\mathcal{G} \to g^{-1}(x)\mathcal{G}(x,y,t)\;\; g(y)$$

when A^o_μ , $F^o_{\mu\nu}$ are transformed following eq.(1.8), (1.9), where g belongs to the representation of the gauge group G for which D^o_μ is the appropriate covariant derivative, for us the adjoint representation. The analogue of Tr (in eq.(1.35)) is

$$\int d^4x\; d^4y\;\; \delta(x-y)\; \mathrm{tr}\,\mathcal{G}(x,y;t) = \mathrm{tr}\int d^4x\, \mathcal{G}(x,x;t) \tag{1.38}$$

where the residual trace is over group indices (or, sometimes, Lorentz indices) and corresponds to a discrete sum. The definition

$$\zeta_{-D^2}(s) = \frac{1}{\Gamma(s)}\int_0^\infty dt\; t^{s-1}\int d^4x\; \mathrm{tr}\,\mathcal{G}(x,x;t) \tag{1.39}$$

provides an analytic function of s with poles at s = 1,2 but regular at s = 0 so that $\zeta_{-D^2}(0)$, $\zeta_{-D^2}(0)$ are well defined and calculable. (Actually that is not quite true because of infrared divergences. However, these are in principle removable by working on a 4-sphere of radius a and letting $a \to \infty$ at the end of the calculation. Some details of this are given in ref.[22] but will be ignored here.)

We note from scaling the operator A in eqs.(1.30)(1.31a,b) that

$$\zeta_{A/\lambda}(s) = \lambda^s\; \zeta(s)$$

and hence that

$$\zeta'_{A/\lambda}(0) = \ln\lambda\; \zeta_A(0) + \zeta'_A(0) \tag{1.40}$$

a relationship which clearly generalises to the operator case also. This tells us, in terms of $\zeta(0)$, the scaling behaviour of any determinant. In particular, for the determinants occurring in eq.(1.29) we can immediately investigate the result of scaling μ once the various $\zeta(0)$ are known. The scaling behaviour is of interest in its own right because it illustrates the way the quantum field theory violates scale invariance, (and indeed conformal invariance [4,21]) to leave the Euclidean analogue of a Poincaré invariant theory.

1.3 Computation of $\zeta(0)$

We shall consider, for definiteness, the operator $-(D^o)^2$ and write, following numerous authors [16,24,25], an asymptotic expansion for $\mathcal{G}(x,y;t)$ near $t = 0$;

$$\mathcal{G}(x,y;t) = \frac{1}{16\pi^2 t^2} e^{-\frac{|x-y|^2}{4t}} \Sigma\, a_n(x,y)\, t^n. \tag{1.41}$$

We deduce immediately that

$$\zeta_{-D^2}(0) = \frac{1}{16\pi^2}\int d^4x \;\; \mathrm{tr}\; a_2(x,x)$$

Since the factor $1/\Gamma(s)$ removes all but the contribution from the pole at $s = 0$ in the integral in eq.(1.39). Furthermore, $\mathcal{G}(x,y;t)$ is supposed to satisfy eqs.(1.36) and (1.37) and substituting the expression (1.41) in the latter yields a recurrence relation for $a_n(x,y)$;

$$n\, a_n(x,y) + (x-y)_\mu D_\mu\, a_n(x,y) = D^2\, a_{n-1}(x,y) \tag{1.42a}$$

$$(x-y)_\mu D_\mu\, a_o(x,y) = 0 \tag{1.42b}$$

$$a_o(x,x) = 1 \tag{1.42c}$$

Eqs.(1.42) are easily solved iteratively in $(x-y)$:

$$a_2(x,x) = \frac{1}{12} F_{\mu\nu} F_{\mu\nu} \tag{1.43}$$

and hence,

$$\zeta_{-D^2}(0) = \frac{1}{192\pi^2}\int d^4x\; \mathrm{tr}\; F_{\mu\nu}F_{\mu\nu} = -\frac{C(A)\cdot k}{6}, \tag{1.44}$$

using eq.(1.11) and the fact that the two traces are performed over different representations of the group. (The number $C(A)$ depends on the representation and is defined relative to the fundamental representation as follows. In the fundamental representation we normalise the generators λ^a so that $\mathrm{Tr}\; \lambda^a\lambda^b = \frac{1}{2}\delta^{ab}$, and other representations are constructed by decomposing tensor products of these. For other representations T^a, $\mathrm{tr}\; T^aT^b = C(T)\;\delta^{ab}$ and, in particular for the adjoint representation of SU(N) or Sp(N) we would have $C(A) = N$ or $N + 1$, respectively).

A similar computation for the operator

$$(\Delta_1)_{\mu\nu} = -\;\delta_{\mu\nu} D^2 - 2[F_{\mu\nu},\;]$$

can easily be performed except that it is necessary to subtract out the zero modes first (otherwise eq.(1.39) makes no sense at all). We may do this by inserting a projection operator and defining

$$\mathcal{G}'_{\mu\nu}(x,y;t) = \int d^4z\; \mathcal{G}_{\mu\rho}\;(x,z;t)\,[\delta(z-y)\delta_{\rho\nu} - \pi_{\rho\nu}(z,y)] \tag{1.45}$$

and manipulate the ζ function replacing $\mathcal{G}$ by $\mathcal{G}'$. The result is,

$$\begin{aligned}\zeta_{\Delta_1}(0) &= \frac{1}{16\pi^2}\int d^4x\; \mathrm{tr}\; a_{2_{\mu\mu}}(x,x) - \int d^4x\; \mathrm{tr}\,\pi(x,x)\\ &= \frac{-5}{48\pi^2}\int d^4x\; \mathrm{tr}\; F_{\mu\nu}F_{\mu\nu} - N(k)\\ &= \frac{10}{3}\; C(A)k - N(k)\end{aligned} \tag{1.46}$$

where N(k) is the number of zero modes of Δ_1. Notice that these two calculations already tell us (because of eq.1.28) what N(k) is, i.e.

$$-N(k) + \frac{10}{3} C(A)k = - \frac{4C(A)k}{6}$$

$$N(k) = 4C(A)k \tag{1.47}$$

the correct result [26,8], and a check on our arithmetic. (This is however only true for sufficiently large k; if $k < \frac{1}{2}N$ for SU(N) or $k < N$ for Sp(N) the counting is more complicated [8]). Moreover we now have enough information to discuss the effect of scaling μ in eq.(1.29). Let $\mu = \tilde{\mu}/\lambda$, then from eqs.(1.40), (1.44), (1.46) we have

$$\mu^{N(k)} \frac{\det(-D^2/\mu^2)}{[\det(\Delta_1/\mu^2)]^{\frac{1}{2}}} = \tilde{\mu}^{N(k)} \frac{\det -D^2/\tilde{\mu}^2}{[\det \Delta_1/\mu^2]^{\frac{1}{2}}} \lambda^{-N(k)-\zeta_{\Delta_1}(0)+2\zeta_{-D^2}(0)} \tag{1.48}$$

and the exponents of λ collect together to give $-\frac{11}{3} C(A)k$. The dependence of eq. (1.48) on λ can be incorporated into a redefinition of the coupling constant

$$g^{-N(k)} e^{-\frac{8\pi^2 k}{g^2}} \lambda^{-\frac{11}{3} C(A)k} = \frac{1}{\tilde{g}^{N(k)}} e^{-\frac{8\pi^2 k}{\tilde{g}^2}}$$

where $\tilde{g}$ obeys the equation

$$\frac{d\tilde{g}}{d\ln\lambda} = \frac{-\frac{\tilde{g}^3}{16\pi^2}\frac{11}{3}C(A)}{1 - \frac{\tilde{g}^2 C(A)}{4\pi^2}} , \tag{1.49}$$

in agreement with the usual β function to lowest order in $\tilde{g}$ [27] as we expected it should.

1.4 Computation of det $(-D^2)$

There does not appear to be any routine way of computing the determinants occurring as factors in eq.(1.29), either by computing $\zeta'_{-D^2}(0)$ directly which involves a greater knowledge of the heat function $\mathcal{G}(x,y;t)$ than we can muster or by computing the eigenvalues of D^2 and using the original expression eq.(1.32) for $\zeta_{-D^2}(s)$, which amounts to the same thing. When k = 1, but not otherwise, this latter method can be used and this corresponds to the original direct calculation of 'T HOOFT [15]. It is a special case for other reasons. For example, the parameter space is an orbit of the conformal group [28], a property not shared by the solutions with given k > 2. For k = 1, this allows a computation of the determinant by considering its conformal variations along the lines proposed by YONEYA [21] and by FROLOV and SCHWARZ [21,4], without needing any detailed knowledge of eigenvalues. For general k it seems profitable to follow refs.[17,20,22] and attempt to set up a differential equation for $\zeta'_{-D^2}(0)$ with respect to variations of any of the instanton parameters, not merely those corresponding to conformal transformations.

The starting point is eq.(1.39) which we write formally as

$$\zeta_{-D^2}(s) = \frac{1}{\Gamma(s)} \int_0^\infty dt\, t^{s-1} \operatorname{tr}(e^{D^2 t}) . \tag{1.50}$$

We recall that the residue of the pole at s = 0 in the function represented by the integral in eq.(1.50) is proportional to the classical action which does not vary under small changes in the instanton parameters,

$$\delta \int d^4x \operatorname{tr} a_2(x,x) \quad 0 .$$

Hence $\delta\zeta_{-D^2}$ is of order s near s = 0,

$$\delta\zeta_{-D^2}(s) = s \left[\int_0^\infty dt\, t^{s-1} \operatorname{tr} \delta e^{tD^2} \right]_{s=0} + O(s^2) . \tag{1.51}$$

In other words $\delta\zeta_{-D^2}(0)$ should be computable and equal to

$$[\int_0^\infty dt \; t^{s-1} \; tr \; \delta e^{tD^2} \;]_{s=0} \tag{1.52}$$

In turn integrating by parts we can re-express $\delta\zeta'_{-D^2}(0)$ as

$$\delta\zeta'_{-D^2}(0) \;=\; [\; -s \int_0^\infty dt \; t^{s-1} \; tr \; e^{D^2 t} \; \delta D^2 \; G \;]_{s=0} \tag{1.53}$$

where G is the inverse of D^2 , the scalar Green function,

$$\vec{D}^2 \; G(x,y) \;=\; G(x,y) \; \overleftarrow{D}^2 \;=-\,\delta(x-y) \; .$$

Finally we obtain:

$$\delta\zeta'_{-D^2}(0) \;=\; \underset{s=0}{\mathrm{res}} \int_0^\infty dt \; t^{s-1} \; tr(e^{tD^2} \; \delta D^2 \; G) \tag{1.54}$$

and, it turns out that the Green functions in an instanton background have a particularly simple form in terms of the ATIYAH, DRINFELD, HITCHIN and MANIN construction [29,30] permitting the calculation of the right hand side of Eq.(1.54) as a finite, manifestly gauge invariant function of the instanton parameters and their small variations. Moreover, at least in the simpler case when D^2 belongs to the fundamental rather than the adjoint representation the variation can be undone. The most complete result obtained so far by BERG and LÜSCHER [31] will be quoted in section 4. For the moment we restrict ourselves to sketching the tricks for evaluating the integrand of eq.(1.54) near t = 0.

If the Green function G(x,y) were not singular as $x \to y$ we could simply use $G(x,y;t) = \delta(x-y) + O(t)$ to evaluate the pole residue in eq.(1.54). However, this is not the case and we have to be more subtle. Since A^o_μ , the background instanton field in terms of which D^2, δD^2, G are all defined, satisfies the classical gauge field equations we can write, following BROWN and CREAMER [32],

$$G(x,y) \;=\; \frac{1}{4\pi^2 (x-y)^2} \; P \exp \int_x^y A.dx \;+\; R(x,y) \tag{1.55}$$

where the path ordered exponential is defined on the straight line path from x to y, and R(x,y) is non-singular when x,y coincide. Then, inserting eq.(1.55) in eq.(1.54) and noting

$$\delta D^2 \;=\; \delta A^o_\mu \; D^o_\mu \;+\; D^o_\mu \; \delta A^o_\mu \tag{1.56}$$

we find that the contribution involving R is simply [32]

$$\int d^4x \;\; Tr \;\; \delta A^o_\mu \; (\vec{D}_\mu \; R(x,y) \;+\; R(x,y) \; \overleftarrow{D}_\mu \;)_{x=y}$$

$$\equiv \int d^4x \;\; Tr \; (\delta A^o_\mu \; J_\mu) \tag{1.57}$$

(for this part we can use $G(x,y;t) = \delta(x-y) + O(t)$). The contribution from the singular part of G is readily found to be zero by a careful consideration of the coefficient of t^o in

$$\int d^4x \; d^4y \;\; Tr \; \{ \textstyle\sum a_n(x,y) \; t^n \; \delta A^o_\mu(y) \; \vec{D}_\mu \; G_s(y,x) \;+\; G_s(x,y) \; \overleftarrow{D}_\mu \; \delta A_\mu(y) \sum a_n(y,x) \; t^n \; \} \; \frac{e^{-\frac{|x-y|^2}{4t}}}{16\pi^2 t^2} \tag{1.58}$$

where

$$G_s(x,y) \;=\; \frac{1}{4\pi^2 (x-y)^2} \;\; P \exp \int_x^y A_\mu dx_\mu \tag{1.59}$$

for details the reader is referred to ref.[22] but the idea is simple enough: expand the integrand as a series in $(x-y)_\mu$ and then integrate over all x to find zero contribution.

To summarise, the ingredients we need to evaluate eq.(1.29) are just δA_μ^o, J and an expansion near x=y of the non-singular part, R(x,y), of the Green function G(x,y), together with some ingenuity in undoing the variation to obtain the determinant. Notice, too, that although the above discussion is really representation invariant we are actually interested in the adjoint representation of the gauge group and it is for that representation that we need the Green function. If the model theory contained scalars or fermions in the fundamental representation of the gauge group all the above arguments would remain valid except that, for fermions, the Dirac operator has eigenvalues of alternating sign and we ought really to consider instead

$$(-i\gamma.D)(i\gamma.D) = D^2 + \sigma_{\mu\nu} F_{\mu\nu} \tag{1.60}$$

(the γ's are the Dirac matrices and $\sigma_{\mu\nu} = \frac{1}{4}[\gamma_\mu,\gamma_\nu]$). The latter has negative eigenvalues (and a zero eigenspace, usually). So we are led to define,

$$\det'(-i\gamma.D) = [\det'(-D^2 - \sigma_{\mu\nu} F_{\mu\nu})]^{\frac{1}{2}} \tag{1.61}$$

2. The ADHM Construction

2.1 The construction of instantons

In this section we should like to focus our attention on the problem of finding all potentials satisfying eq.(1.13) and for which the action S has the value $8\pi^2k$. We shall describe, in elementary terms, the construction of ATIYAH, DRINFELD, HITCHIN and MANIN [6] and how their construction enables us to solve related problems such as finding the Green function G(x,y) for various representations or finding the zero eigenspace of the Dirac operator. A major omission in our simplified discussion will be the proof that the ADHM construction yields all solutions to eq.(1.13); we know of no such argument which does not involve sheaf cohomology in an essential way.

The ADHM construction had its genesis in twistor methods [33]. WARD [34] used these techniques to draw a one-to-one correspondence between self-dual solutions of the Yang-Mills equations and certain holomorphic vector bundles. ATIYAH and WARD [35] developed these ideas further showing that for solutions which are sufficiently well behaved at infinity the construction was necessarily algebraic, reducing the problem from one in complex analysis to one in complex algebraic geometry. (The condition at infinity is that the fields decrease sufficiently fast to enable the solution to be considered on the four sphere S^4, the conformal compactification of Euclidean space R^4). It has since been shown [36] that all finite action solutions satisfy this asymptotic condition. Subsequently, using techniques of algebraic geometry and building on work of BARTH [27] and HORROCKS [38] ADHM obtained their general method for constructing self-dual solutions. We shall follow the notations and conventions of refs.[29] in what follows [39].

To describe the general self-dual solution for a given compact Lie group all we need to do is to describe the general solution for each simple Lie algebra occurring in the decomposition of the Lie algebra of the group and add them together. Quite simple descriptions of the solutions can be given for each of the four sequences of compact simple groups: SU(n+1), O(2n+1), O(2n) or Sp(n), but it suffices to do it for just one sequence because the other groups may be realised as subgroups of suitable groups in that sequence. A similar technique can be used for exceptional groups and, though the details have not been written down, it ought to be straightforward. Of the four main sequences the formalism is simplest for Sp(n) and we shall confine ourselves to this case.

The symplectic groups may be less familiar than the unitary or orthogonal groups. They may be thought of as nxn unitary matrices with quaternion entries. Another way of saying this is that Sp(n) is the group of complex matrices U in SU(2n) which have the property,

$$U^T J U = J \qquad (2.1)$$

where

$$J = \begin{pmatrix} \varepsilon & 0 & 0 & \cdots & 0 \\ 0 & \varepsilon & 0 & \cdots & 0 \\ 0 & 0 & \varepsilon & \cdots & 0 \\ \vdots & \vdots & \vdots & & \vdots \\ 0 & 0 & 0 & \cdots & \varepsilon \end{pmatrix} \qquad (2.2)$$

and the 0's stand for 2x2 zero matrices, ε for the 2x2 matrix

$$\varepsilon = \begin{pmatrix} 0 & 1 \\ -1 & 0 \end{pmatrix} \qquad (2.3)$$

Eq.(2.1) indicates that U can be divided into our nxn array of 2x2 blocks each of the form

$$\begin{pmatrix} y & -\bar{z} \\ z & \bar{y} \end{pmatrix} , \qquad (2.4)$$

i.e. a quaternion, demonstrating the relationship with the previous definition. In particular, we note that Sp(1) is the group of unit quaternions and so the same as SU(2). So, in what follows, to set what is happening in a familiar context we can set n=1 and obtain results for SU(2).

We can write down straight away the result of the ADHM construction for the potential A_μ. It is given by

$$A_\mu = V^\dagger \partial_\mu V \qquad (2.5)$$

where V(x) is an(n+k) x n matrix of quaternion functions of the space-time variable x_μ normalised so that

$$V^\dagger(x)\, V(x) = 1_n . \qquad (2.6)$$

We note that for k=0 the vector potential is a pure gauge because eq.(2.6) informs us that $V(x) \in Sp(n)$. For $k \neq 0$ the quaternion structure of V and eq.(2.6) ensure that the vector potential lies in the Lie algebra of Sp(n) and has the correct hermiticity property, $A_\mu^\dagger = -A_\mu$, for our conventions.

The matrix V(x) is not arbitrary it has to be determined via eq.(2.6) and the following set of linear equation

$$V^\dagger(x)\, \Delta(x) = 0 \qquad (2.7)$$

where $\Delta(x)$ is a (k+n) x k matrix of quaternions, linear in the space time variables x_μ. The latter coordinates may be conveniently incorporated into a quaternion by writing

$$x = x_o - i\underline{x}.\underline{\sigma} = \begin{pmatrix} x_o - ix_3 & -ix_1 - x_2 \\ -ix_1 + x_2 & x_o + ix_3 \end{pmatrix} \qquad (2.8)$$

Thus the elements of the matrix Δ have the structure

$$\Delta_{\lambda i} = a_{\lambda i} + b_{\lambda i}\, x \qquad \begin{matrix} 1 \leqslant \lambda \leqslant n+k \\ 1 \leqslant i \leqslant k \end{matrix} \qquad (2.9)$$

where $a_{\lambda i}$, $b_{\lambda i}$ are quaternions also.

The a's and b's are not arbitrary, however. In order that the potential given by eqs.(2.5)-(2.9) yields a self-dual field strength we also have to have tne following conditions:

$$a^\dagger a,\ b^\dagger b,\ a^\dagger b \text{ are symmetric as kxk quaternion matrices.} \tag{2.10}$$

Equivalently $\Delta^\dagger\Delta$ has to be symmetric for all x. However, since $\Delta^\dagger\Delta$ is manifestly hermitian it must therefore be a real multiple of the unit quaternion matrix $\begin{pmatrix}1&0\\0&1\end{pmatrix}$. We shall show that the conditions (2.10) yield self-dual solutions locally. To ensure that they are non singular we also need that $\Delta^\dagger\Delta$ is non-singular for all x. In particular to ensure finite action we need that $b^\dagger b$ be non-singular. As it recurs repeatedly in calculations we denote the inverse of $\Delta^\dagger\Delta$ by f which is itself a real symmetrical kxk matrix.

We can prove directly that eqs.(2.5)-(2.10) yield a self-dual field strength. For this it is useful to introduce a projection operator P onto the space orthogonal to the columns of Δ. Then

$$P = VV^\dagger = 1 - \Delta f \Delta^\dagger \tag{2.11}$$

so that

$$P^2 = P, \quad P^\dagger = P \tag{2.12}$$

and

$$PV = V, \quad P\Delta = 0. \tag{2.13}$$

By straightforward calculation

$$F_{\mu\nu} = V^\dagger [\partial_\mu P, \partial_\nu P]\, V \tag{2.14}$$

which may be simplified, using eqs.(2.11), (2.7) to show

$$\begin{aligned} V^\dagger \partial_\mu P &= -V^\dagger(\partial_\mu \Delta)\, f\, \Delta^\dagger = -V^\dagger b e_\mu f \Delta^\dagger \\ \partial_\nu P V &= -\Delta f\, e_\nu^\dagger b^\dagger V \end{aligned} \tag{2.15}$$

(where $e_\mu = \partial_\mu x$), to

$$F_{\mu\nu} = V^\dagger b\, (e_\mu f e_\nu^\dagger - e_\nu f e_\mu^\dagger) b^\dagger V. \tag{2.16}$$

Now from eq.(2.8) we may calculate the e_μ and check that

$$e_\mu e_\nu^\dagger - e_\nu e_\mu^\dagger = 2i\, \bar{\eta}^a_{\ \mu\nu}\, \sigma^a \tag{2.17}$$

where $\bar{\eta}^a_{\ \mu\nu}$ is a basis for self-dual tensors [12]

$$\bar{\eta}^a_{\ \mu\nu} = \varepsilon_{oa\mu\nu} - \delta_{a\mu}\,\delta_{o\nu} + \delta_{a\nu}\,\delta_{o\mu}. \tag{2.18}$$

From eqs.(2.16), (2.17) we see that $F_{\mu\nu}$ will be self-dual provided that e_μ commutes with f. This is precisely the condition that $\Delta^\dagger\Delta$ be real, and is guaranteed by eqs. (2.10).

In order to check that the k determining the size of the matrix V and the one appearing in the value of the action, $8\pi^2 k$, are indeed the same it is useful to note the following identities [22,47],

$$\begin{aligned} \tfrac{1}{2}(\mathrm{Tr}(F_{\mu\nu}F_{\mu\nu}) &= \partial^2\, \mathrm{tr}\, (b^\dagger P b f + b^\dagger b f) && (2.19) \\ &= \tfrac{1}{2}\, \partial^2 \partial^2\, \ln \det f && (2.20) \end{aligned}$$

which can be checked using the ingredients described above. Using either of these to write the action as a surface integral, and noting the asymptotic form of $f \sim (b^\dagger b)^{-1} \frac{1}{x^2}$, for large x, yields

$$\int d^4x \, \tfrac{1}{2}\mathrm{tr}\, F_{\mu\nu}F_{\mu\nu} = 8\pi^2 k.$$

Performing the identical procedure but with x replaced by $x^\dagger$ throughout yields all the anti-self-dual solutions.

2.2 Comments on the construction

(a) Gauge transformations

It is interesting to consider the relationship of the construction to gauge transformations. Provided $\Delta^\dagger\Delta$ is non-singular eqs.(2.6), (2.7) determine V up to a transformation of the form

$$V(x) \rightarrow V(x)\; g(x) \tag{2.21}$$

where $g(x) \in Sp(n)$. This induces precisely the gauge transformation eq.(1.8) on the potential A_μ given by eq.(2.5). Thus a given

$$\Delta = a + bx$$

defines a potential up to gauge equivalence. On the other hand different a's and b's may yield the same gauge potential. To see this note that a redefinition

$$a \rightarrow QaK, \qquad b \rightarrow QbK \tag{2.22}$$

for any $Q \in Sp(n+k)$, $K \in GL(k,\mathbb{R})$, leads to $\Delta \rightarrow Q\Delta K$ and

$$V \rightarrow QV$$

yields a solution of the modified equations (2.7) but leaves A_μ unchanged. In fact, it follows from the work of ATIYAH et al [18] that different a's and b's will yield the same gauge equivalence class of potentials if and only if they are related by a transformation like eq.(2.22). Thus a and b are gauge invariant parameters for the solution with the gauge equivalence classes of the solutions in one-to-one correspondence with the orbits of the manifold of solutions of the quadratic constraints (2.10) under the group of transformations (2.22). In a sense the infinte dimensional gauge group has been traded for a finite dimensional (though non-compact) group of transformations.

(b) Conformal transformations [29,22]

The effect of conformal transformations on the space-time coordinates x can be easily expressed in terms of the construction. The conformal transformations on four-dimensional Euclidean space take the form $x \rightarrow x'$ with

$$x' = (\alpha x+\beta)\,(\gamma x+\chi)^{-1} \tag{2.23}$$

where α, β, γ, χ are quaternions. The conformal invariance of the Yang-Mills action means that if $A_\mu(x)$ is a solution to the classical equations so also is

$$A_\nu'(x) = A_\mu(x')\,\frac{\partial x'_\mu}{\partial x_\nu} \tag{2.24}$$

If $A_\mu(x)$ is given by eq.(2.5) then so also is $A_\mu'(x)$ with

$$V'(x) = V(x'(x)) \tag{2.25}$$

which can be obtained from

$$\Delta'(x) = a' + b'x$$

with

$$a' = a\chi + b\beta\ , \quad b' = a\gamma + b\alpha \tag{2.26}$$

(c) Special forms and the number of instanton parameters

If we wish we can choose the transformation K in eq.(2.22) so that

$$b^\dagger b = 1_k \tag{2.27}$$

the unit kxk matrix. Then Q can be selected so that b assumes the form

$$b = \begin{pmatrix} 0 \\ 1_k \end{pmatrix} \tag{2.28}$$

where the first n rows are identically zero. The freedom then remaining is eq.(2.22) with,

$$Q = \begin{pmatrix} U & 0 \\ 0 & R \end{pmatrix} \quad K = R^{-1} \qquad U\varepsilon Sp(n), \quad R\varepsilon O(k). \tag{2.29}$$

If we write

$$a = \begin{pmatrix} \xi \\ \eta \end{pmatrix} \tag{2.30}$$

where ξ, η are nxk and kxk matrices of quaternions, respectively, the constraint eqs.(2.10) take the form that η and $\xi^\dagger\xi + \eta^\dagger\eta$ be symmetric. These constraints leave $4(n+1)k + \frac{1}{2}k(k-1)$ degress of freedom. The O(k) part of the transformations (2.29) remove a further $\frac{1}{2}k(k-1)$ parameters. Finally, provided $n \leq k$ the Sp(n) part in eq. (2.29) reduces the number down to [8,26]

$$4(n+1)k - n(2n+1) \tag{2.31}$$

On the other hand, if $n > k$ we can use the Sp(n) to further refine the special form (eq.2.30) reducing the first (n-k) rows to zero. There remains an Sp(k) group preserving this canonical form and we see we have

$$4(k+1)k - k(2k+1) = k(2k+3) \tag{2.32}$$

parameters in this case [8]. So, a k instanton solution can always be obtained from one in an Sp(k) subgroup of Sp(n) when $k < n$.

(d) The 'T HOOFT solutions

The class of solutions first found by 'T HOOFT[40] and conformally extended by JACKIW, NOHL and REBBI [28] is easily found for n=1 (i.e. SU(2) remember). Let us take

$$\begin{aligned} a_{oi} &= \alpha_o\lambda_i\ , \quad a_{ij} = \lambda_o\alpha_i\ \delta_{ij} \qquad\qquad 1 \leq i,\ j \leq\ k \\ b_{oi} &= -\lambda_i \qquad\ b_{ij} = -\lambda_o\ \delta_{ij} \end{aligned} \tag{2.33}$$

where all the λ's are real, the α's quaternions. Then it is easy to see that the constraints (2.10) are satisfied and a solution to eq.(2.7) is provided by

$$V_\sigma = -\ (x - \alpha_\sigma)^{\dagger -1}\ \lambda_\sigma\ \phi^{-\frac{1}{2}} \qquad\qquad \sigma = 0,\ 1,\ \ldots,\ k \tag{2.34}$$

where

$$\phi = \sum_0^k \frac{\lambda_\sigma^2}{|x-\alpha_\sigma|^2}\ . \tag{2.35}$$

Without any loss of generality we can write,

$$A_\mu dx_\mu = \tfrac{1}{2}\,(V^\dagger\, dV - dV^\dagger\, V), \tag{2.36}$$

and substitute in from eq.(2.34) to deduce,

$$A_\mu = \tfrac{1}{2}\, i\, \eta^a_{\mu\nu}\, \sigma^a\, \partial_\nu\, \ell n\phi \tag{2.37}$$

where $\eta^a_{\mu\nu}$ is a basis for anti-self-dual tensors, i.e.

$$\eta^a_{\mu\nu} = \varepsilon_{oa\mu\nu} + \delta_{a\mu}\,\delta_{o\nu} - \delta_{a\nu}\,\delta_{o\mu} \tag{2.38}$$

analogous to $\bar{\eta}^a_{\mu\nu}$ [12]. Eq.(2.37) has the form of the well-known ansatz [41], corresponding to the lowest ansatz of the ATIYAH-WARD programme [35].

3. Green functions [42,29,30]

In the previous section we reviewed many of the properties of the ADHM construction useful for the programme of evaluating functional integrals. Here we shall see how to find Green functions in various situations.

3.1 Covariant differentiation and projection operators

As a preliminary we wish to say a little more about the geometrical nature of the covariant derivative in the ADHM construction.

Consider a field transforming under the fundamental representation of the gauge group (2n dimensional for Sp(n)) and its covariant derivative in terms of the instanton potential

$$D_\mu\phi = \partial_\mu\phi + V^\dagger\partial_\mu V\,\phi = V^\dagger\partial_\mu(V\phi). \tag{3.1}$$

This suggests that we define

$$\hat{\phi} = V\phi \quad \text{and} \quad D_\mu\hat{\phi} = \widehat{D_\mu\phi} \tag{3.2}$$

in which case

$$D_\mu\hat{\phi} = P\partial_\mu\hat{\phi} \tag{3.3}$$

We may describe these results as follows. The matrix V maps the 2n dimensional complex field ϕ with a 2(n+k) dimensional complex field $\hat{\phi}$ which lies in a variable 2n dimensional subspace of $\mathbb{C}^{2(n+k)}$, the subspace

$$E_x = \{\, \xi \;:\; P(x)\xi = \xi \,\} \tag{3.4}$$

orthogonal to $\Delta(x)$ onto which P is the projection operator. The collection of spaces $\{E_x\}$ as x varies over Euclidean space $\mathbb{R}^4$ (or S^4 after performing conformal compactification)forms a vector bundle. Inside this vector bundle of subspaces of $\mathbb{C}^{2(n+k)}$, covariant differentiation is just defined by ordinary differentiation followed by projection.

3.2 Fundamental representation Green function

There is a remarkably simple result for the Green function of a scalar particle (i.e. of the gauge covariant laplacian) transforming under the fundamental representation of the gauge group, in the background field of a given instanton. The Green function is defined by

$$D^2\, G(x,y) = -\delta(x-y) \tag{3.5}$$

with the boundary condition $G(x,y) \to 0$ as $|x| \to \infty$.

The Green function for the ordinary rather than the covariant Laplacian (on $\mathbb{R}^4$) is defined by

$$\partial^2 G_o(x,y) = -\delta(x-y) \tag{3.6}$$

and the same boundary condition and hence given by

$$G_o(x,y) = \frac{1}{4\pi^2 |x-y|^2} \tag{3.7}$$

The solution to eq.(3.5) is

$$G(x,y) = \frac{V^\dagger(x)\, V(y)}{4\pi^2\, |x-y|^2} \tag{3.8}$$

and, in fact, supplies the simplest generalisation of eq.(3.7) transforming correctly under the gauge group, viz.

$$G(x,y) \rightarrow g(x)^{-1}\, G(x,y)\, g(y) \tag{3.9}$$

under a transformation of the form of eqs.(1.8) (1.9) and sec.(2.2a).

Clearly as $x \sim y$, $G(x,y) \sim G_o(x,y)$ so, if we can show

$$D^2\, G(x,y) = 0 \qquad x \neq y \tag{3.10}$$

the result (3.8) will be proved. From (3.1) we can see that

$$D^2 \left(\frac{V^\dagger(x)\, V(y)}{|x-y|^2} \right) = V^\dagger(x)\, \partial_\alpha \left(P(x)\, \partial_\alpha \left[\frac{P(x)\, V(y)}{|x-y|^2} \right] \right)$$

$$= V(x)^\dagger \left\{ \partial^2 P(x) + \partial_\alpha P(x)\, \partial_\alpha P(x) - \frac{4(x-y)_\alpha\, \partial_\alpha P(x)}{|x-y|^2} \right\} \frac{V(y)}{|x-y|^2} \tag{3.11}$$

and at first sight looks unlikely to be zero because of the different singularities as $x \rightarrow y$. However, noting that

$$V(x)^\dagger\, \partial_\alpha P(x)\, V(y) = -\, V(x)^\dagger\, b\, e_\alpha\, f(x)\, (\Delta(x)^\dagger - \Delta(y)^\dagger)\, V(y)$$

$$= -\, V(x)^\dagger\, b\, e_\alpha\, f(x)\, (x^\dagger - y^\dagger)\, b^\dagger\, V(y) \tag{3.12}$$

we may rewrite the right hand side of eq.(3.11) as

$$V(x)^\dagger \left\{ \partial_\alpha (P\, \partial_\alpha P) + 4\, b\, f(x)\, b^\dagger \right\} \frac{V(y)}{|x-y|^2} \tag{3.13}$$

It is quite straightforward, using the machinery of section 2, to demonstrate that

$$P\, \partial_\alpha (P\, \partial_\alpha P) + 4Pbf\, b^\dagger = 0 \tag{3.14}$$

and the result (3.8) for the Green function (which is unique) is verified [29,30].

Because of the special relationships between the eigenfunctions of the appropriate generalisations of the Laplace operator for fields of various (spatial) spins and also of the Dirac operator [42], varying the spin of the Green function presents little difficulty. On the other hand, changing the representation of the gauge group from the fundamental respresentation discussed above, to say the adjoint representation, presents more trouble [42]. However, the problems have to be solved since it is precisely results in the adjoint representation of the group that are important for the computation of determinants discussed in section (1). The following two subsections are devoted to these topics.

3.3 Solutions of the massless Dirac equation

The ADHM construction involves the space of solutions to the massless Dirac equation and these can be reconstructed easily in the formalism we have developed [29,43]. Take as a representation of the γ matrices

$$\gamma_\alpha = \begin{pmatrix} 0 & e_\alpha \\ e^\dagger_\alpha & 0 \end{pmatrix}, \quad \gamma_5 = \gamma_0\gamma_1\gamma_2\gamma_3 = \begin{pmatrix} -1 & 0 \\ 0 & 1 \end{pmatrix} \tag{3.15}$$

The Dirac equation can be decomposed conveniently into positive and negative chirality parts :

$$\psi = \begin{pmatrix} \psi_L \\ \psi_R \end{pmatrix}, \quad e_\alpha D_\alpha \psi_R = 0, \quad e^\dagger_\alpha D_\alpha \psi_L = 0. \tag{3.16}$$

Since

$$e_\beta D_\beta e^\dagger_\alpha D_\alpha = D^2 + \frac{i}{2}\bar{\eta}_{\alpha\beta}F_{\alpha\beta} \tag{3.17}$$

whilst

$$e^\dagger_\beta D_\beta e_\alpha D_\alpha = D^2 + \frac{i}{2}\eta_{\alpha\beta}F_{\alpha\beta} = D^2 \tag{3.18}$$

(since $F_{\alpha\beta}$ is self-dual, $\eta_{\alpha\beta}$ is anti-self-dual), we see that $D^2\psi_R = 0$. The latter has no normalisable solutions and hence all solutions to the Dirac equation must have negative chirality. The index theorem tells us that there are precisely k independent solutions of negative chirality which we have to find [44]. We may regard the field ψ_L as a 2nx2 dimensional matrix, separating out the group and spinor indices. In which case the second of equations (3.16) reads

$$V^\dagger \partial_\alpha (V\psi\, \bar{e}_\alpha) = 0 \tag{3.19}$$

(the - denotes ordinary conjugation of the 2x2 matrices representing the components of e_α). If we multiply the eq.(3.19) by V on the left and ε on the right we can rewrite it as

$$P\, \partial_\alpha (\,\Psi\, e_\alpha) = 0 \tag{3.20}$$

where

$$\Psi = V\psi\varepsilon.$$

The kind of algebra that lead to a proof of eq.(3.14) leads (as a stop on the way actually) also to

$$P\, \partial_\alpha (\, Pbfe_\alpha\,) \equiv 0. \tag{3.21}$$

Hence uniqueness leads us to the result;

$$(\psi_2)^{\,i}_{\,a,\,I} = (V^\dagger bf\varepsilon)_{a,iI} \tag{3.22}$$

where $1 \leqslant a \leqslant 2n$, $I = 1, 2$ solves the Dirac equation for each fixed i, $1 \leqslant i \leqslant k$.

3.4 The adjoint representation and tensor products [45]

If we use q to denote the fundamental representation of a gauge group G we can obtain the adjoint representation by decomposing $q \otimes \bar{q}$; that is we regard a field in the adjoint representation as a two index object, one index transforming according to the fundamental representation and the other its complex conjugate. Then the covariant derivative

$$D_\mu = 1 \otimes 1\, \partial_\mu + A_\mu \otimes 1 + 1 \otimes \bar{A}_\mu, \tag{3.23}$$

in an obvious notation, is the appropriate one.

The adjoint scalar Green function is again defined by eq.(3.5) but with D_μ defined by eq.(3.23) instead of eq.(3.1). Optimistically the simple result (3.8) for the fundamental representation would lead to a guess for the adjoint representation (or rather for the tensor product $q \otimes \bar{q}$) as follows:

$$G(x,y) = \frac{V(x)^\dagger V(y) \otimes \overline{V^\dagger(x) V(y)}}{4\pi^2 |x-y|^2} \tag{3.24}$$

However, as first pointed out by BROWN et al. for the 'T HOOFT case, this is not correct and another term non-singular as $x \to y$ has to be added [42].

To understand where this term comes from we prefer to think about the problem in a more general context. Consider the direct product $G_1 \times G_2$ of simple groups G_1, G_2 and suppose we have instanton solutions

$$A_1 = V_1^\dagger \partial_\mu V_1 , \qquad A_2 = V_2^\dagger \partial_\mu V_2 , \tag{3.25}$$

described in the ADHM way, for each group. Consider also a field transforming under the fundamental representation of each. Then its covariant derivative is obtained from,

$$D_\mu = 1 \otimes 1 \, \partial_\mu + A_{1\mu} \otimes 1 + 1 \otimes A_{2\mu} \tag{3.26}$$

and we would regain the special case of interest, the adjoint representation of G, by taking $G_1 = G_2 = G$ and $A_{2\mu} = A_{1\mu} = A_\mu$. We call the solution (3.25) for $G_1 \times G_2$ the tensor product of the solutions for G_1 and G_2.

The problem of finding the Green function for a tensor product, that is of solving eq.(3.5) with D_μ defined by (eq.3.23), should have a simple solution for the following reason. The solution for $G_1 \times G_2$ will yield a solution for any group G containing $G_1 \times G_2$; for example, if we choose G to be a group of matrices containing the tensor products of matrices in G_1 and G_2 as a subgroup. In particular we can consider $Sp(n_1) \times Sp(n_2) \subset SU(4n_1n_2)$, taking the tensor products of the $2n_1$ and $2n_2$ dimensional matrices of $Sp(n_1)$ and $Sp(n_2)$ to yield $4n_1n_2$ dimensional matrices. Then a field transforming under the tensor product of the fundamental representations of G_1 and G_2 can be regarded as transforming under the fundamental representation of G. Now, it has to be possible to describe the potential

$$A_\mu = A_{1\mu} \otimes 1 + 1 \otimes A_{2\mu} \tag{3.27}$$

within the ADHM construction since it is supposed to yield every solution for G. Thus it must be possible to represent A_μ as $\tilde{V}^\dagger \partial_\mu \tilde{V}$ for a suitable matrix $\tilde{V}$ and furthermore, the Green function is simply given by

$$\frac{\tilde{V}^\dagger(x) \, \tilde{V}(y)}{4\pi^2 |x-y|^2}$$

looked at this way the problem is to locate the potential A_μ (eq.3.27) on the list provided by the ADHM construction. This is not as straightforward as it sounds, however, because it is not enough simply to find a $\tilde{V}$ such that

$$\tilde{V}^\dagger \partial \tilde{V} = V_1^\dagger \partial V_1 \otimes 1 + 1 \otimes V_2^\dagger \partial V_2 \tag{3.28}$$

for, if it were, taking

$$\tilde{V} = V_1 \otimes V_2 \tag{3.29}$$

would lead to the incorrect formula (3.24) for the Green function.

Apart from a direct check that (3.24) and (3.29) do not work we can see that they cannot be quite right by a dimensional argument. For Sp(n) we saw in section 2 that V was a 2(n+k) x 2n matrix where k is the instanton number; for SU(N), V would be (N+2k) x N and k has the same interpretation. For the potential (3.27) we can compute the instanton number k in terms of k_1, n_1 and k_2, n_2 viz.

$$k = 2n_1k_2 + 2n_2k_1 \tag{3.30}$$

for $Sp(n_1) \times Sp(n_2) \subset SU(4n_1n_2)$. Thus $\tilde{V}$ should be a $(4n_1n_2 + 4n_1k_2 + 4n_2k_1) \times 4n_1n_2$ dimensional matrix. On the other hand $V_1 \otimes V_2$ is a $4(n_1+k_1)(n_2+k_2) \times 4n_1n_2$ dimensional matrix. The necessary reduction in effective dimensions is achieved [45] by multiplying by a matrix with $4k_1k_2$ zero eigenvalues. Then

$$G(x,y) = \frac{1}{4\pi^2 |x-y|^2} V_1(x)^\dagger \otimes V_2^\dagger(x) (1 - \mathfrak{M}) V_1(y) \otimes V_2(y) \tag{3.31}$$

which means that, effectively,

$$\tilde{V}(x) = (1 - \mathfrak{M})^{\frac{1}{2}} V_1(x) \otimes V_2(x)$$

instead of (3.29). It turns out (for details on all this see ref.[45]) that

$$\tilde{V}(x)^\dagger \tilde{V}(y) = V_1^\dagger(x) V_1(y) \otimes V_2^\dagger(x) V_2(y) + O(|x-y|^2) \tag{3.32}$$

and, differentiating with respect to y and setting x = y we observe that $\tilde{V}^\dagger \partial V$ does actually yield the correct potential A_μ (eq.3.27).

The matrix $\mathfrak{M}$ has many remarkable properties not the least of which is conformal invariance and is expressed entirely as a function of a_1, b_1; a_2, b_2 the parameter bearing matrices of the ADHM construction for the potentials $A_{1\mu}$, $A_{2\mu}$, respectively. $\mathfrak{M}$ is complicated to write out in a suffix notation so the reader is referred to ref.[45] for details, including a pictorial notation which, in our view, elucidates the algebra greatly.

To understand where the $\mathfrak{M}$ comes from mathematically we need to rephrase the ADHM construction more abstractly [46]. Really we are going backwards into its original formulation. Let us rewrite the information contained in Δ in a rather different way. Suppose

$$\theta = \begin{pmatrix} \pi \\ \omega \end{pmatrix} \tag{3.33}$$

is a complex four-vector, π, ω being complex two-vectors, and suppose $(w_i) \in \mathbb{C}^k$. Then we can form $A_\theta(w)$ an element of the 2(k+n) dimension complex space into which Δ maps, regarding it as k+n two-vectors

$$(A_\theta(w))_\lambda = (a_{\lambda i}\, \pi + b_{\lambda i}\, \omega)\, w_i . \tag{3.34}$$

For each $\theta \in C^4$ A_θ provides a map from a k-dimensional complex space W to a 2(k+n) dimensional complex space V. To see what properties of A correspond to the conditions on Δ define an antisymmetric form on V by

$$[\, v_1, v_2 \,] = v_1^T J\, v_2 \tag{3.35}$$

where J is the antisymmetric matrix defined in (eq.2.2). Then it is easy to check that provided a, b are matrices of quaternions, the conditions on them, eq.(2.10), are equivalent to

$$[\, A_\theta(w^{(1)}), A_\theta(w^{(2)}) \,] = 0 \quad \forall \; w^{(1)}, w^{(2)} \in W,\ \theta \in C^4. \tag{3.36}$$

We may express the fact that a, b are composed of quaternions in the following way. Define antilinear maps as follows

$$\sigma: W \to W, \qquad \sigma(w)_i = \bar{w}_i \tag{3.37}$$

$$\sigma: V \to V, \qquad \sigma(v)_\lambda = \varepsilon \bar{v}_\lambda \tag{3.38}$$

$$j: C^4 \to C^4 \qquad j\theta = \begin{pmatrix} \varepsilon & \bar{\pi} \\ \varepsilon & \bar{\omega} \end{pmatrix}, \tag{3.39}$$

then the condition that a, b be matrices of quaternions is

$$\sigma A_\theta(w) = A_{j\theta}(\sigma w). \tag{3.40}$$

Now, we may rephrase eq.(3.36) further. Consider the dual space W^* of mappings $W \to C$. Defining $A_\theta : V \to W^*$ by

$$A_\theta(v)\ (w) = [\ v, A_\theta(w)\ ] \tag{3.41}$$

we find that eq.(3.36) is now equivalent to

$$A_\theta^{\ 2} = 0 \tag{3.42}$$

If we also define,

$$\sigma: W \to W^* \qquad \sigma(w^*)\ (w) = \overline{w^*(\sigma(w))} \tag{3.43}$$

we have

$$\sigma A_\theta(v) = A_{j\theta}(\sigma v). \tag{3.44}$$

It is convenient to collect together all the rephrasing and new information by defining

$$V_{-1} = W,\ V_o = V,\ V_1 = W^*,\ V_r = 0 \ \text{ if } \ |r| > 1 \tag{3.45}$$

and extending A_θ to a map $V_r \to V_{r+1}$ such that

$$A_\theta(V_r) = 0 \ \text{ if } \ |r| > 1. \tag{3.46}$$

The grade of ξ, $\{\xi\}$, is defined to be r if and only if $\xi \varepsilon V_r$ and we may extend the definition of the bracket operation, eq.3.35, by

$$[\xi,\eta] = 0 \qquad \text{if} \qquad \{\xi\} + \{\eta\} \neq 0 \tag{3.47}$$

$$[w,w^*] = [w^*,w] = w^*(w) \qquad \text{if} \qquad w \varepsilon W,\ w^* \varepsilon W^*. \tag{3.48}$$

Thus, finally we have a sequence of spaces V_r and maps $A_\theta : V_r \to V_{r+1}$, a form [,] and antilinear maps $V_r \to V_r$ such that the following properties are true:

(i) $[\xi,\eta] = 0 \quad \text{if} \quad \{\xi\} + \{\eta\} \neq 0$

(ii) $[\xi,\eta] = -(-)^{\{\xi\}\{\eta\}}\ [\eta,\xi]$

(iii) $A_\theta^{\ 2} = 0$

(iv) $[A_\theta(\xi),\eta] = (-)^{\{\xi\}}\ [\xi,A_\theta(\eta)]$

(v) $[\sigma\xi,\sigma\eta] = \overline{[\xi,\eta]}$

(vi) $\sigma^2(\xi) = -(-)^{\{\xi\}}\ \xi$

(vii) $\sigma A_\theta(\xi) = A_{j\theta}(\sigma\xi)$

(viii) $\langle\xi,\eta\rangle = [\sigma\xi,\eta]$ is positive definite if $\xi,\eta \ \varepsilon\ V_o$

(ix) $V_r = 0, \quad |r| > 1.$

We can call such a set of axioms an instanton complex for a symplectic group. To do the same for a unitary group the axioms have to be modified slightly. Instead of (ii) we ought to write

$$\text{(ii)}' \qquad [\xi,\eta] = (-1)^{\{\xi\}\{\eta\}} [\eta,\xi]$$

and, instead of (ii):

$$\text{(vi)}' \qquad \sigma^2(\xi) = (-1)^{\{\xi\}} \xi.$$

Having set up all this machinery we can take another look at tensor products. Given two symplectic instanton complexes $A_\theta^{(i)} : V_r^{(i)} \to V_{r+1}^{(i)}$ etc., $i = 1,2$, we can define a product $\tilde{A}_\theta : \tilde{V}_r \to \tilde{V}_{r+1}$ etc., in the following way. Let

$$\tilde{V}_r = \bigoplus_s V_s^{(1)} \otimes V_{r-s}^{(2)} \tag{3.49}$$

(note that $\tilde{V}_r = 0$ only if $|r| > 2$, so this complex will be bigger),

$$\tilde{A}_\theta(\xi_1 \otimes \xi_2) = A_\theta^{(1)}(\xi_1) \otimes \xi_2 + (-)^{\{\xi_1\}} \xi_1 \otimes A_\theta^{(2)} (\xi_2) \tag{3.50}$$

$$[\xi_1 \otimes \xi_2, \eta_1 \otimes \eta_2] = (-)^{\{\xi_1\}\{\xi_2\}} [\xi_1,\eta_1][\xi_2,\eta_2] \tag{3.51}$$

and

$$\sigma(\xi_1 \otimes \xi_2) = \sigma(\xi_1) \otimes \sigma(\xi_2) \tag{3.52}$$

(note $\{\xi_1 \otimes \xi_2\} = \{\xi_1\} + \{\xi_2\}$).

With the new definitions (3.49) - (3.52) we can check straightforwardly that the unitary group axioms hold except for (viii) and (ix). The problem is that the complex is now too big. To cut it down to size we need a way to remove $\tilde{V}_{\pm 2}$ and this we can manage in the following way. Define $V'_{-2} = \tilde{V}_{-2}$ and V'_r to be the subspace of $\tilde{V}_r$ spanned by vectors of the form $A_\theta(\xi)$, $\xi \varepsilon V'_{r-1}$ ($\theta \varepsilon \mathbb{C}^4$) and consider the orthogonal spaces

$$V_r = \{ \xi \varepsilon \tilde{V}_r : [\xi,\eta] = 0 \ \forall \eta \varepsilon V'_{-r} \}. \tag{3.53}$$

It can be shown that $A_\theta = \tilde{A}_\theta$ maps $V_r \to V_{r+1}$, σ maps $V_r \to V_r$ and all the axioms hold for maps A_θ, σ applied to V_r. The appropriate V, W for the tensor product are subspaces of

$$\tilde{V}_0 = W^{(1)} \otimes W^{(2)*} \oplus V^{(1)} \otimes V^{(2)} \oplus W^{(1)*} \otimes W^{(2)} \tag{3.54}$$

and

$$\tilde{V}_{-1} = W^{(1)} \otimes V^{(2)} \oplus V^{(1)} \otimes W^{(2)}. \tag{3.55}$$

The operator $1 - \mathfrak{M}$ which has been inserted into eq.(3.31) is just the part of the orthogonal projection operator onto V_o which maps $V^{(1)} \otimes V^{(2)}$ onto $V^{(1)} \otimes V^{(2)}$. The interested reader is urged to consult ref.[46] for more details on all this.

Once we have discovered the way to write tensor products within the same formalism as the fundamental representation we are able to transcribe results computed in the fundamental representation (for Green functions, determinants, etc.,) to higher representations.

4. Calculation of $\det(-D^2)$

In this section we should like to outline briefly the steps towards computing the expression (1.57), at least in the fundamental representation of G so that the algebra is kept under control. We need δA_μ^o and R(x,y) and these can be computed via the formula for the instanton potential described in section (2), eqs.(2.5) - (2.10). Clearly, from eqs.(2.7) and (2.6) we must have

$$\delta V^{\dagger}\Delta + V^{\dagger}\delta\Delta = 0 \tag{4.1}$$

$$\delta V^{\dagger}V + V^{\dagger}\delta V = 0 \tag{4.2}$$

and hence that

$$\delta V^{\dagger} = - V^{\dagger}\,\delta\Delta\,(\Delta^{\dagger}\Delta)^{-1}\,\Delta^{\dagger} + \delta u\, V^{\dagger} \tag{4.3}$$

$$\delta u + \delta u^{\dagger} = 0 \tag{4.4}$$

In other words, from (2.5)

$$\begin{aligned}\delta A^{o}_{\mu} &= - V^{\dagger}\delta\Delta\, f\,\Delta^{\dagger}\partial_{\mu}V - V^{\dagger}\partial_{\mu}\,(\Delta f\,\delta\Delta^{\dagger}V)\\ &\qquad + \delta u\, V^{\dagger}\partial_{\mu}V + V^{\dagger}\partial_{\mu}V\,\delta u^{\dagger} + \partial_{\mu}\delta u^{\dagger}\\ &= V^{\dagger}\delta\Delta f\partial_{\mu}\Delta^{\dagger}\,V - V^{\dagger}\partial_{\mu}\,\Delta f\,\delta\Delta^{\dagger}V + D^{o}_{\mu}\,\delta u^{\dagger}.\end{aligned} \tag{4.5}$$

The final term of eq.(4.5) contains $\delta u^{\dagger}$, is an infinitesimal gauge,transformation and may be ignored.

A computation of R(x,y) is slightly more involved, but since we know the Green function all we have to do is compute the path ordered exponential to the order in $(x-y)_{\mu}$ that we require. (For details see ref[22]). This is perhaps most easily done in the following way. A moment's thought suggests the following form

$$\Phi = P\exp\int_{x}^{y} A.dx = V^{\dagger}(x)\,[1 + |x-y|^{2}\, bH(x,y)b^{\dagger}]\,V(y) \tag{4.6}$$

where H(x,y) can be determined using the defining differential equation for Φ:

$$(x-y)_{\mu}\,\vec{D}_{\mu}\,\Phi = 0 = (x-y)_{\mu}\,\Phi\,\overleftarrow{D}_{\mu}.$$

Thus, setting $\bar{x} = \tfrac{1}{2}(x+y)$

$$H(x,y) = \tfrac{1}{2}\, f(\bar{x}) + \frac{1}{12}\, f(\bar{x})[(x-y)\,\Delta^{\dagger}(\bar{x})\,b - b^{\dagger}\Delta(\bar{x})(x-y)^{\dagger}]\, f(\bar{x}) + 0(|x-y|^{2}) \tag{4.7}$$

and we may compute J_{μ}:

$$J_{\mu} = \frac{1}{12\pi^{2}}\, V^{\dagger}bf(e_{\mu}\,\Delta^{\dagger}b - b^{\dagger}\Delta e^{\dagger}_{\mu})fb^{\dagger}V \tag{4.8}$$

Note also

$$D^{o}_{\mu}\,J_{\mu} = 0 \tag{4.9}$$

by direct calculation. Eq.(4.9) guarantees that the final term in eq.(4.5) plays no role in the expression for $\delta\zeta_{-D^2}(0)$, eq.(1.56).

Putting together the ingredients we have assembled gives

$$\mathrm{Tr}(\delta A^{o}_{\mu}J_{\mu}) = \frac{1}{12\pi^{2}}\,\mathrm{tr}[Pbf\,(e_{\mu}\Delta^{\dagger}b-b^{\dagger}\Delta e^{\dagger}_{\mu})fb^{\dagger}P(\delta\Delta f\partial_{\mu}\Delta^{\dagger}-\partial_{\mu}\Delta f\delta\Delta^{\dagger})] \tag{4.10}$$

where P is the projection operator $VV^{\dagger}$ (eqs.(2.11) - (2.13)). Finally using eq.(3.14) and

$$P\delta P = - P\delta\Delta f\Delta^{\dagger},$$

We find, after some algebra, the expression [20,22]

$$\mathrm{Tr}\,(\delta A^{o}_{\mu} J_{\mu}) = \frac{1}{96\pi^2}\, \mathrm{tr}(\delta P[\partial^2 P(1-P)\, \partial^2 P - 2\partial_\mu P\, \partial^2 P\, \partial_\mu P]). \tag{4.11}$$

The latter expression is rather nice because it is manifestly gauge invariant, has a <u>finite</u> integral over R^4 and depends solely on the basic building block, P, of the ADHM construction.

In order to find an expression for the determinant itself the variation with respect to the instanton parameters has to be undone. Some progress towards this was made recently by OSBORN [47]. However, the most complete results in this direction have been obtained by BERG and LÜSCHER [31] who managed to rewrite $\delta\zeta_{-D^2}(0)$ as the total variation of four and five dimensional integrals in the following way,

$$\mathrm{Tr}\,(J_\mu \delta A_\mu) = \frac{1}{48\pi^2}\, \delta[\, 20\, \mathrm{tr}\, (b^\dagger b f b^\dagger b f) - \mathrm{tr}\, (f\partial_\mu f^{-1}\, f\partial_\mu f^{-1}\, f\partial_\nu f^{-1}\, f\partial_\nu f^{-1}) + \theta\,] + \partial_\mu X_\mu \tag{4.12}$$

where θ is given by

$$\theta = \frac{1}{5}\int_0^1 dt\; \mathrm{tr}(m_\alpha m_\beta m_\gamma m_\delta m_\sigma)\; \varepsilon_{\alpha\beta\gamma\delta\sigma}$$

$$m_\alpha = m^{-1}\partial_\alpha m$$

$$m = (1+x^2)(1-t) + t\, \Delta^\dagger \Delta. \tag{4.13}$$

An outstanding problem is to find a set of variables in terms of which these integrals may be performed. A possible way of defining such a set (analogous to the 'positions' of the 't Hooft instantons) has been suggested by FATEEV <u>et al</u> [20] but so far their conjectures have not been substantiated.

However it has proved possible to generalise the work of YONEYA [21] and FROLOV and SCHWARZ [4,21] to compute the conformally non-invariant part of the determinant of $-D^2$. To do this the variations δP to be considered are those due to conformal changes of parameters and may be completely integrated up . For details we refer the reader to ref.[22] and quote the result

$$\delta\zeta_{-D^2}(0) = \delta(\frac{1}{12k}\, \ell n\, \det\, \{b^\dagger b \otimes b^\dagger b\, M\}\,) \tag{4.14}$$

where M is the conformal invariant matrix forming a major ingredient of the operator $\mathfrak{M}$ [30,45], eq.(3.31), and defined by

$$[b^\dagger b \otimes a^\dagger a + a^\dagger a \otimes b^\dagger b - \mathrm{tr}(a^\dagger b \otimes a^\dagger b)]\, M = 1 \otimes 1. \tag{4.15}$$

The trace appearing in eq.(4.15) is over the quaternion indices only so M is a $k^2 \times k^2$ matrix whose components are all real.

Finally, a word about infra red divergences. By considering the conformal properties of the operator $-D^2 + R/6 = \mathfrak{D}$ and applying the results to the conformal scaling of the metric induced by changing the radius, a, of the four dimensional sphere (on which we ought to be working) we find that the leading dependence on a (for large a) of $\det \mathfrak{D}/\mu^2$ is simply a factor $(\mu a)^{N/45}$ where N is the dimension of the gauge field representation [22]. Hence, the variation of $\ell n \det \mathfrak{D}/\mu^2$ with respect to changes in the instanton parameters is independent of the radius a as $a \to \infty$ and is therefore finite. This fact we have already noted by inspection of the integral over $\mathbb{R}^4$ of eq. (4.11).

5. <u>A more transparent model in two dimensions</u>

As mentioned in the last section, it has not yet proved possible to evaluate the multiple instanton contribution to the gauge theory functional integral so any discussion of physical implications is of necessity speculative. In two dimensions, however, the so-called CP^{n-1} models [12] bear, at least superficially, a resemblance to gauge theories in four dimensions, and, furthermore, enjoy the additional property of calculability. As an illustration, therefore, we would like to sketch the CP^1 calculation (of FROLOV and SCHWARZ [48], BERG and LÜSCHER [13]) using the techniques described in section 1. We shall follow the path forged by FROLOV and SCHWARZ. They pioneered the use of the ζ function definition of determinants in this context and we think it is the clearest way to proceed.

5.1 CP^{n-1} models in two dimensions

CP^{n-1} theories can be thought of as described by an (n^2-1) component real field $N^i(x)$ $i = 1,2,\ldots.n^2-1$, (i.e. transforming under the adjoint representation of SU(n)) satisfying subsidiary conditions which force it

(a) to have constant length

(b) to lie in an SU(n) orbit with little group U(n-1).

Then, if the field N^i obeys the boundary condition $\underline{N} \to$ constant at two dimensional Euclidean infinity (which is necessary for finite action) we can regard $\underline{N}(x)$ as a mapping from the two dimensional sphere S^2 (compactified Euclidean space) to the SU(n) orbit in which the field lies. These maps fall into homotopy classes since [49]

$$\Pi_2\ (SU(n)/\ U(n-1)\ ) = \Pi_1\ (U(n-1)\ ) = \mathbb{Z} \tag{5.1}$$

and the instantons are field configurations representing the equivalence classes under homotopy. (Note $CP^{n-1} \cong SU(n)/U(n-1)$, hence the name).

More analytically, we note that if the n-component complex field ψ transforms under the fundamental representation of SU(n) then setting $N^i \propto \psi^\dagger\lambda^i\psi$, where λ^i are the generators of the Lie algebra of SU(n) in the fundamental representation, and $\psi^\dagger\psi = 1$ automatically implies conditions (a) and (b) above. The appropriate action is, in terms of ψ :

$$S = \tfrac{1}{2} \int d^2x\ \ (D_\mu\psi)^\dagger . D_\mu\psi \tag{5.2}$$

where

$$D_\mu = \partial_\mu - \psi^\dagger . \partial_\mu\psi \tag{5.3}$$

Clearly, the action (5.2) is invariant under $\psi \to e^{i\Lambda(x)}\psi$. Furthermore, we may note that

$$S \geqslant \tfrac{1}{2} \int d^2x\ (D_\mu\psi)^\dagger\ \varepsilon_{\mu\nu}\ D_\nu\psi = 4\pi|k| \qquad k = 0,\pm1,\pm2\ldots \tag{5.4}$$

with equality if

$$D_\mu\psi = \pm i\ \varepsilon_{\mu\nu}\ D_\nu\psi \tag{5.5}$$

which is the two dimensional analogue of self duality in the gauge theory case. The integral on the right hand side of eq.(5.4) provides the analytic way of computing the homotopy class of the field N.

Unlike the gauge theory case, where the solution of the self-dual equation required considerable sophistication, here eq.(5.5) is easily solved by setting [50]

$$Z = x_o + ix, \quad \text{and} \quad \psi = \psi_o(1,\ \underline{\hat{\psi}}) \tag{5.6}$$

Then eq.(5.5) becomes simply (for the + sign)

$$\frac{\partial}{\partial \bar{z}} \cdot \hat{\psi} = 0, \quad |\psi_o|^2 = \frac{1}{1+|\hat{\psi}|^2} \tag{5.7}$$

so that $\hat{\psi}$ is a function of z only, not $\bar{z}$. The most general possibility, appropriate to instanton number k and satisfying the boundary conditions, is that each component of $\hat{\psi}$ has the form

$$\hat{\psi}_i = c_i \frac{\prod_{j=1}^{k} (z - a_{ij})}{\prod_{j=1}^{k} (\bar{z} - a_{oj})}, \quad i = 1\ldots.n-1, \tag{5.8}$$

revealing at a glance all the instanton degrees of freedom.

For n=2 $\hat{\psi}$ is a single component and the instanton has 2k+1 complex degress of freedom, $a_{1i} \equiv a_i$, $a_{oi} \equiv b_i$, c_i $i = 1 \ldots.k$.

The remarkable result of refs.[48,13] is that, in this case, the leading contribution of the k-instanton to the functional integral defining the theory, eq.(5.41) below, is the partition function (evaluated at unit temperature, in suitable units) for a classical neutral Coulomb gas of 2k particles each of mass m (the renormalisation group invariant mass, see below), k of which are positively charged the other k negatively charged. It is curious that the result reveals classical statistics but perhaps that is an artifact of the leading approximation and will be corrected (to Fermi or Bose?) when higher orders are taken into account. One of the important facts about the two dimensional Coulomb gas is that T = 1 is a critical point at which the pressure diverges [51] indicating that a dilute (i.e. non-interacting or weakly interacting) gas approximation is incorrect and a poor approximation to the actual behaviour of the field theory. However, despite the divergence in the pressure (and hence the instanton density) the Green functions of the theory may still have a sensible limit as T → 1. It is not yet known if any of this has a counterpart in the four dimensional gauge theory but there has been some speculation about it recently (see BELAVIN et al [20]).

5.2 Estimation of the function integral in the CP^1 case.

For n = 1 the Euclidean Green functions of the theory are defined by

$$Z = \frac{1}{Z_o} \int d[N] \; \Phi(\underline{N}) \; e^{-\frac{1}{g^2}S(N)} \tag{5.9}$$

where N has three real components and the measure d[N] is to be thought of as

$$d[N] = \prod_x [\, d^3N \, \delta(\underline{N}^2 - 1) \,],$$

a rather formal expression perhaps best thought of via a lattice approximation. As in section (1) we shall sum over contributions coming from quadratic approximation of S near each of its local minima. For this purpose it is useful to change variables to $\hat{\psi}$ (eq.(5.6)) and herafter drop the ^ i.e.,

$$N_1 + iN_2 = \frac{2\psi}{1+|\psi|^2}, \quad N_3 = \frac{1-|\psi|^2}{1+|\psi|^2} \tag{5.10}$$

$$\tfrac{1}{2}(\partial_\mu \underline{N})^2 = 2\,\frac{\partial_\mu \bar{\psi} \partial_\mu \psi}{(1+|\psi|^2)^2} = \frac{4}{(1+|\psi|^2)^2} \left[\, \left|\frac{\partial\psi}{\partial z}\right|^2 + \left|\frac{\partial\psi}{\partial \bar{z}}\right|^2 \,\right]. \tag{5.11}$$

Thus, near the k-instanton configuration

$$S = 4\pi k + 8 \int d^2x \; \frac{\left|\frac{\partial\psi}{\partial \bar{z}}\right|^2}{(1+|\psi|^2)^2} \tag{5.12}$$

whilst for the measure

$$\underset{x}{\pi}\,(d^3N\,\delta(\underline{N}^2-1)) \;=\; \underset{x}{\pi}\,\frac{d^2\psi}{(1+|\psi|^2)^2} \;\equiv\; d[\psi] \tag{5.13}$$

and eq.(5.9) becomes

$$Z = \sum_k Z_k \;=\; \frac{1}{Z_o}\sum_k \int d[\Psi]\;\;\Phi\;\exp\left[-\frac{4\pi k}{g^2} + \frac{8}{g^2}\int d^2x\,\frac{|\frac{\partial\psi}{\partial z}|^2}{(1+|\psi|)^2}\right]. \tag{5.14}$$

We can now proceed as in section (1) apart from one important remark. The presence of the extra factor $(1+|\psi|^2)^{-2}$ in the measure, eq.(5.13), needs careful consideration when we come to evaluate N, (eq.1.5), the determinant of the zero mode normalisation matrix. Normally, in the absence of such a factor we would write

$$N_{ij} \;=\; \int d^2x\;\overline{\frac{\partial\psi_o}{\partial p_i}}\;\frac{\partial\psi_o}{\partial p_j} \tag{5.15}$$

where ψ_o is the background instanton field (i.e. eq.(5.8)) and the p_i $i = 1\ldots 2k+1$ collectively label the instanton complex degrees of freedom. However, in this case, the inner product in eq.(5.15) is incorrect (and indeed the integral not well-defined) and ought to be replaced by

$$N_{ij} \;=\; \int d^2x\;\frac{1}{(1+|\psi_o|^2)^2}\;\overline{\frac{\partial\psi_o}{\partial p_i}}\;\frac{\partial\psi_o}{\partial p_j}\;. \tag{5.16}$$

To convince oneself of eq.(5.16) it is necessary to perform a careful limiting process from an integral defined on a discrete lattice to the functional integral.

In order to compute detN we rewrite ψ_o explicitly as

$$\psi_o \;=\; C\,\frac{\overset{k}{\underset{1}{\pi}}(z-a_i)}{\overset{k}{\underset{1}{\pi}}(z-b_i)} \;\equiv\; \frac{n}{d} \tag{5.17}$$

where n and d are two polynomials of degree k. Let

$$d^2\,\frac{\partial\psi_o}{\partial p_j} \;=\; \sum_{\ell=1}^{2k+1} P_{j\ell}\,z^\ell,$$

a polynomial of degree 2k+1, then a simple direct calculation reveals

$$\det P \;=\; (\text{constant})\; c^{2k}\;\underset{i<j}{\pi}\,(b_i-b_j)\;\underset{i<j}{\pi}\,(a_i-a_j)\;\underset{i,j}{\pi}\,(a_i-b_j) \tag{5.18}$$

and the zero mode determinant is given by,

$$\det N \;=\; |\det P|^2\;\det M \tag{5.19}$$

where

$$M_{ij} \;=\; \int d^2x\;\frac{(\bar z)^i\,z^j}{(|n|^2+|d|^2)^2}\;. \tag{5.20}$$

Fortunately, we do not have to compute detM since this factor will cancel exactly with a piece arising from the functional integral over the variations away from the instanton, to which we now turn.

A useful change of variables in the functional integral (5.14), bearing in mind eq.(5.13) and the fact that we are only interested in the leading contribution as $g \to 0$, is to set;

$$\psi = \psi_o + \frac{(|d|^2+|n|^2)}{d^2} \chi \equiv \psi_o + \frac{\rho}{d^2} \chi . \tag{5.21}$$

Then, the awkward measure, eq.(5.13) becomes [in the limit $g \to 0$] simply $\prod_x d^2\chi$ and the differential operator whose determinant we need to compute is just

$$Q = \rho \frac{\partial}{\partial z} \frac{1}{\rho^2} \frac{\partial}{\partial \bar{z}} \rho . \tag{5.22}$$

That Q is the appropriate operator follows straightforwardly from the exponent in eq.(5.14) using eq.(5.21), remembering that d is a function of z alone and integrating by parts.

Finally, inserting the factors of μ, π, g in the appropriate way we arrive at the leading order contribution to Z_k

$$Z_k = \frac{1}{Z_o} \left(\frac{8\mu^2}{\pi g^2} \right)^{N(k)} \frac{1}{(k!)^2} \int \prod_i d^2p_i \; (\det N) \; \Phi \; e^{-\frac{4\pi k}{g^2}} \; [\det{}' \{-Q/\mu^2\}]^{-1} . \tag{5.23}$$

(The factors of k! are to compensate for multiple counting if we just went blindly ahead and integrated over all values of a and b. This happens because ψ_o is invariant under any permutation of the a's or b's amongst themselves).

5.3 Computation of $\det'(-Q/\mu^2)$.

We proceed as in section 1 to use the ζ function technique to define the determinant by first setting up a differential equation for its variation with respect to a small change in any of the instanton parameters p_i (a_i or b_i) $i = 1....k$. Thus starting from

$$\zeta_Q(s) = \frac{1}{\Gamma(s)} \int_0^\infty dt \; t^{s-1} \; \mathrm{tr} \{ e^{Qt} (1 - \Pi) \} \tag{5.24}$$

(remembering to project out all the zero modes), using

$$\delta Q = \delta(\rho \frac{\partial}{\partial z} \rho^{-2} \frac{\partial}{\partial \bar{z}} \rho) = (\delta \ln\rho) Q + Q(\delta \ln\rho) = 2\rho \frac{\partial}{\partial z} \rho^{-1} (\delta \ln\rho) + \rho^{-1} \frac{\partial}{\partial \bar{z}} \rho \tag{5.25}$$

and noting

$$Q \rho \frac{\partial}{\partial z} \rho^{-1} = \rho \frac{\partial}{\partial z} \rho^{-1} \; \rho^{-1} \frac{\partial}{\partial \bar{z}} \rho^2 \frac{\partial}{\partial z} \rho \equiv \rho \frac{\partial}{\partial z} \rho^{-1} \hat{Q} \tag{5.26}$$

we find

$$\delta\zeta_Q(s) = \frac{1}{\Gamma(s)} \int_0^\infty dt \; t^s \; \mathrm{tr} \, (\delta Q \, e^{Qt})$$

$$= \frac{1}{\Gamma(s)} \int_0^\infty dt \; t^s \; 2 \frac{d}{dt} \; \mathrm{tr} \{ \delta \ln\rho \, (e^{Qt}(1 - \Pi) - e^{\hat{Q}t}) \} . \tag{5.27}$$

Integrating the latter by parts gives

$$\delta\zeta_Q(s) = - \frac{2s}{\Gamma(s)} \int_0^\infty dt \; t^{s-1} \; \mathrm{tr} \{ \delta \ln\rho \, (e^{Qt}(1 - \Pi) - e^{\hat{Q}t}) \}, \tag{5.28}$$

where we have noted that $\hat{Q}$ has no zero modes.

Eq.(5.28) is very useful since we may immediately deduce that the complete determinant, or rather its variation, is given by a calculable pole residue, i.e.

$$\delta\zeta'_Q(0) = -2 \operatorname{Res}_{s=0} \int_0^\infty dt\, t^{s-1}\, \mathrm{tr}\, \{ \delta \ell n\rho\, (e^{Qt}(1 - \Pi) - e^{\hat{Q}t}). \qquad (5.29)$$

for the ζ function itself evaluated at $s = 0$,

$$\zeta_Q(0) = \operatorname{Res}_{s=0} \int_0^\infty dt\, t^{s-1}\, \mathrm{tr}\, \{ e^{Qt}(1 - \Pi)\} \qquad (5.30)$$

Following the two dimensional version of the discussion in section (1.3) we write

$$e^{Qt} = \mathcal{G}(z,y;t) = \frac{e^{-\frac{|z-y|^2}{t}}}{\pi t} \quad \Sigma\, a_n\, (z,y) t^n \qquad (5.31)$$

$$e^{\hat{Q}t} = \hat{\mathcal{G}}(z,y;t) = \frac{e^{-\frac{|z-y|^2}{t}}}{\pi t} \quad \Sigma\, \hat{a}_n (z,y) t^n \qquad (5.32)$$

and calculate a_1, $\hat{a}_1$. After some straightforward algebra we find

$$a_1(z,z) = - \hat{a}_1(z,z) = \frac{\partial}{\partial \bar{z}} \frac{\partial}{\partial z} \ell n\rho \qquad (5.33)$$

Hence, substituting a_1, $\hat{a}_1$ into eqs.(5.30) and (5.29) yields

$$\zeta_Q(0) = \frac{1}{\pi} \int d^2x\, \tfrac{1}{4}\, \partial^2 \ell n\rho \;-\; N(k)$$
$$= k - N(k) \qquad (5.34)$$

and

$$\delta\zeta'_Q(0) = -\frac{1}{\pi} \int d^2x\, \delta\ell n\rho\, \partial^2 \ell n\rho \;+\; 2 \int d^2x\, \delta\ell n\rho\, \Pi(z,z)$$
$$= \delta I_1 + \delta I_2 \qquad (5.35)$$

The integral δI_2 is relatively straightforward to compute and provides the piece to cancel detM, (eq.5.19). We note that from eqs. (5.16) and (5.20) the zero mode projection operator has the form,

$$\pi(z,y) = \sum_{i,j} z^i \frac{(M^{-1})_{ij}}{\rho^2} \bar{y}^j \qquad (5.36)$$

and

$$\delta I_2 = 2 \int d^2x\, \frac{\delta\rho}{\rho^3} \sum_{i,j} z^i (M^{-1})_{ij}\, \bar{z}^j \qquad (5.37)$$

Using eq.(5.20) again we may recognise eq.(5.37) to be

$$\delta I_2 = - \delta\, \mathrm{tr}\, \ell n M = - \delta\, \ell n\, \det M \qquad (5.38)$$

as claimed. The other integral δI, is a little trickier to deal with so we shall omit the details and quote the result,

$$\delta I_1 = - \delta\, [\, 2k\, \ell n|c|^2 + 2 \sum_{i,j} \ell n\, |a_i - b_j|^2 \,]. \qquad (5.39)$$

Hence, undoing the variations is now trivial and, combining eqs. (5.39), (5.38) and (5.18) we obtain (aside from a constant factor)

$$\det N\,[\det{}'(-Q/\mu^2)]^{-1} = (\mu^2)^{k-N(k)} \exp\{-\sum_{i,j} \ln|a_i-b_j|^2 + \sum_{i<j} \ln|a_i-a_j|^2 + \sum_{i<j} \ln|b_i-b_j|^2\}. \qquad (5.40)$$

(Notice that the dependence on c has disappeared; a more correct version of the above calculation working on the sphere S^2 yields an overall factor $\frac{1}{(1+|c|^2)^2}$ together with a logarithmic divergence in the (large) radius of the sphere. However, the latter is independent of k and so will cancel between the numerator and denominator of eq. (5.9). Our intention here was to give a simple outline of the calculation rather than supply all details. For these the reader is referred to refs.[48,13].

Finally, we have deduced

$$Z_k \sim \left(\frac{8}{\pi g^2}\right)^{2k+1} \mu^{2k}\, e^{-\frac{4\pi k}{g^2}} \frac{1}{(k!)^2} \int \prod_i d^2a_i\, d^2b_i d^2c\, \Phi(\underline{a},\underline{b},c)\, e^{-V(a,b)} \qquad (5.41)$$

and again, we note that a scaling of μ can be compensated by a redefinition of g^2, in terms of the scale parameter, corresponding to a $\beta(g)$ agreeing with previous calculation to leading order. Indeed, apart from an overall factor of g^{-2} the contribution of μ and g can be replaced by m^{2k} and we recognise the partition function for a neutral system of 2k particles of mass m and opposite with charges interacting via the two dimensional Coulomb potential V(a,b) and evaluated at the (singular) point T = 1.

What we have just described seems to us to be a remarkable and beautiful phenomenon, but it remains to be seen whether anything similar occurs in four dimensional gauge theories. Optimists think that it will, of course, but whatever the outcome of the instanton calculations we feel they will be interesting and perhaps unexpected.

Acknowledgements

We are grateful to Hugh Osborn for sharing his understanding with us on many occasions.

REFERENCES (This list has no pretensions to being either extensive or chronological)

1. C.N. Yang and R. Mills Phys. Rev. 96 (1954) 191.
2. H. Eichenherr Nucl. Phys. B146 (1978) 215.
   I.V. Golo and A. Perelomov Phys. Lett. 79B (1978) 112.
3. E.S. Abers and B.W. Lee Physics Reports C9 (1973) 1.
4. A more careful consideration of the points mentioned in this section can be found in A.S. Schwarz, Lett. Math. Phys. 2 (1978) 201, 247; Commun. Math. Phys. 64 (1979) 233; Commun. Math. Phys. 67 (1979) 1.
5. A.A. Belavin, A.M. Polyakov, A.S. Schwarz and Yu.S. Tyupkin, Phys. Letts. 59B (1975) 85.
6. M.F. Atiyah, V.G. Drinfeld, N.J. Hitchin and Yu.I. Manin, Phys. Letts. 65A (1978) 185.
7. V.G. Drinfeld and Yu.I. Manin, Commun. Math. Phys. 63 (1978) 177.
8. A.S. Schwarz, Phys. Letts. 67B (1977) 189.
   C.W. Bernard, N.H. Christ, A.H. Guth and E.J. Weinberg, Phys. Rev. D16 (1977) 2967.
   M.F. Atiyah, N.J. Hitchin and I.M. Singer, Proc. Nat. Acad. Sci. 74 (1977) 262.
9. For a review see ref.3, L. Fadeev, Teor. Mat. Fiz 1 (1969) 3.
10. D. Amati and A. Rouet, Phys. Letts. 73B (1978) 39.
11. I.M. Singer, seminar of University of Irvine (February, 1979).
12. G. 't Hooft, Phys. Rev. Letts. 37 (1976) 8; Phys. Rev. D14 (1976) 343; erratum Phys. Rev. D18 (1978) 2199.
13. B. Berg and M. Lüscher, Commun. Math. Phys. 69 (1979) 57.
14. F.T. Ore Phys. Rev. D16 (1977) 2577.
    A.A. Belavin and A.M. Polyakov Nucl. Phys. B123 (1977) 429.
    Yu.A. Bashilov and S. Pokrovsky, Nucl. Phys. B143 (1978) 431.
    G.M. Shore Cambridge preprint DAMTP 78/19 to appear in Ann. Phys.
    C. Bernard, Phys. Rev. D19 (1979) 3013.
15. S. Chadha, A. D'Adda, P. Di Vecchia and F. Nicodemi, Phys. Letts. 72B (1977) 103.
16. J.S. Schwinger, Phys. Rev. 82 (1951) 664.
    B.S. DeWitt, Dynamical theory of groups and fields, Gordon and Breach (1965); Phys. Reports 19C (1975) 295.
17. D.B. Ray and I.M. Singer, Adv. Math. 7 (1971) 145.
18. J.S. Dowker and R. Critchley Phys. Rev. D13 (1976) 3224; Phys. Rev. D16 (1977) 3390.
19. S. Hawking, Commun. Math. Phys. 55 (1977) 133.
20. A.A. Belavin, V.A. Fateev, A.S. Schwarz and Yu.S. Tyupkin, Phys. Letts. 83B (1979) 317.
21. But see also: T. Yoneya Phys. Letts. 71B (1977) 407.
    I.V. Frolov and A.S. Schwarz, Phys. Letts. 80B 406.
22. E. Corrigan, P. Goddard, H. Osborn, S. Templeton CalTech preprint 68-726 May, 1979 to appear in Nucl. Physics B.
23. Bateman Manuscript Project (McGraw-Hill 1953) Vol.1 p.32.
24. P.B. Gilkey, The index theorem and the heat equation, Publish or Perish (1974).
25. R.T. Seeley, Proc. Symp. Pure Math. 10 (1971) 288.
26. R. Jackiw and C. Rebbi, Phys. Letts. 67B (1977) 189.
27. For a review see H.D. Politzer, Phys. Reports C14 (1974) 129.
28. R. Jackiw, C. Nohl and C. Rebbi, Phys. Rev. D15 (1977) 1642.
29. E. Corrigan, P. Goddard, D.B. Fairlie and S. Templeton, Nucl. Phys. B140 (1978) 31.
30. N.H. Christ, E. Weinberg and N.K. Stanton, Phys. Rev. D18 (1978) 2013.
31. B. Berg and M. Lüscher, DESY preprint August 1979, to appear in Nuclear Physics.
32. L.S. Brown and D.B. Creamer, Phys. Rev. D18 (1978) 3695.
33. R. Penrose, in Quantum gravity, eds. C.J. Isham, R. Penrose and D.W. Sciama (Oxford University press, 1975).
34. R.S. Ward, Phys. Lett. 61A (1977) 81.
35. M.F. Atiyah and R.S. Ward, Commun. Math. Phys. 55 (1977) 117.
36. K. Uhlenbeck, Bull. Amer. Math. Soc. 1 (1979) 579.
37. W. Barth, Math. Ann 226 (1977) 125.
38. G. Horrocks, Proc. Lond. Math. Soc. (3) 14 (1964) 689.

39. But see also J. Madore, J.L. Richard and R. Stora, Physics Reports C49 (1979) 113.
40. G. 't Hooft, unpublished.
41. F. Wilczek, in Quantum confinement and field theory, ed. D. Stump and D. Weingarten (New York, John Wiley and Sons, 1977).
E. Corrigan and D.B. Fairlie, Phys. Letts. 67B (1977) 69.
42. L.S. Brown, R.D. Carlitz, D.B. Creamer and C. Lee, Phys. Letts 70B (1977) 180 or 71B (1977) 103; Phys. Rev. D17 (1978) 1583.
43. H. Osborn, Nucl. Phys. B140 (1978) 45.
44. See e.g. M.F. Atiyah, N.J. Hitchin and I.M. Singer, Proc. Roy. Soc. A362 (1978) 425.
45. E. Corrigan, P. Goddard, S. Templeton, Nucl. Phys. B151 (1979) 93.
46. V.G. Drinfeld and Yu.I. Manin, Yad. Fiz. 29 (1979) 1646.
47. H. Osborn, Cambridge preprint DAMTP 79/12, to appear in Nuclear Physics B.
58. I.V. Frolov and A.S. Schwarz, J.E.T.P. Letts. 28 (1978) 249.
V.A. Fateev, I.V. Frolov and A.S. Schwarz, Nucl. Phys. B154 (1979) 1. (This paper also contains an analysis of the relationship between Pauli-Villars and proper time regularisation).
49. See, for example, D. Husemaller, Fibre Bundles, McGraw-Hill (1966)
50. This method stems from A.A. Belavin and A.M. Polyakov, J.E.T.P. Letters 22 (1975) 245.
51. J. Fröhlich, Commun. Math. Phys. 47 (1976) 233.
52. M. Lüscher, Phys. Lett. 78B (1978) 465.
A. D'Adda, P. Di Vecchia and M. Lüscher, Nucl. Phys. B146 (1978) 63; Nucl. Phys. B152 (1979) 125.

The Interpretation of Higgs Fields as Yang Mills Fields

David B. Fairlie
Department of Mathematics
University of Durham
Durham, U.K.

One of the greatest challenges for the particle theorist at the present time is to try to discover a more general principle lying behind the phenomenologically successful Weinberg-Salam (1) theory of weak interactions. The central issue is to elucidate the nature of the Higgs fields. In W-S they play a dual role - to give a mass to the charged and neutral vector mesons - and also, through their interaction with the fermions to give a mass to the leptons in the theory. Now their role as fundamental scalars is being increasingly challenged - Hawking and his collaborators (2) have recently claimed that, with appropriate boundary conditions a scalar particle will acquire a mass equal to the planck mass on account of interactions with gravitational instantons over a distance comparable with the planck length - spinning particles are immune from this effect, and thus one school of thought which regard Higgs scalars as composites of lepto-quarks is exempt from this criticism, provided the composite is sufficiently large. This school has not yet to my knowledge constructed a satisfactory proto-theory in terms of pre-quarks without primitive Higgs fields which gives a convincing reduction to a phenomenological theory equivalent to W-S.

I am going to talk about a second interpretation of Higgs fields which has grown from developments in the solution of classical Yang Mills equations, namely, that in which the Higgs fields masquerade as Yang Mills gauge fields. I should mention at the outset that this is by no means a final solution which I am going to present, but rather a few possible connections which may dovetail into a complete theory. The initial motivation springs from the monopole theory of 'tHooft and Polyakov (3). Consider a time independent solution of 4 dimensional Yang Mills theory. Then

$$F_{io} = [D_i, A_o] = D_i A_o$$
$$F_{ij} = [D_i, D_j] \tag{1}$$

where D_i is the usual covariant derivative $\partial_i + [A_i, \ ]$

The action for this theory is

$$\frac{1}{4}\int \left(F_{ij}{}^2 + (D_i A_o)^2\right) d^4x \tag{2}$$

Factoring out the contribution of the time integral gives a theory which is identical to that of a 3 dimensional Yang Mills theory coupled to a Higgs potential A_o, the so called Prasad-Sommerfield limit in which there is no self interaction of the Higgs field. This interpretation was well known to many workers in monopole theory, and has been exploited particularly by David Olive. The equivalence holds not only at the level of the action, but also at that of the equations of motion also. In attempting to find solutions of this theory, the constraint $\mathrm{Tr}\, A_o{}^2 \to$ const as $\underline{x} \to \infty$ is imposed as a remnant of the Higgs potential.

It would be advantageous to try to exploit the same mechanism to give a deeper understanding of W-S theory; as I hope to show attempts along this line raise as many questions as they answer - yet I believe there is an underlying germ of truth in this approach. Three problems must be overcome - the Higgs field must transform as a doublet, and not as the adjoint representation of U(2), the theory must be time dependent, and a self interaction term must be generated. The simplest way of satisfying all three criteria is to extend both the dimensions of the underlying space time manifold and also of the internal symmetry group, and to attempt to recognise the W-S theory as a restriction of this case. Consider a Yang Mills theory in six dimensions and look for a special solution of the form

$$\begin{pmatrix} gA_\mu - g'B_\mu I & 0 \\ 0 & \lambda g'B_\mu \end{pmatrix} \quad \mu = 1 \,..\, 4 \tag{3}$$

where A_μ transforms under SU(2) while B_μ transforms under U(1) and

$$A_5 = g\begin{pmatrix} 0 & i\phi \\ -i\phi^\pm & 0 \end{pmatrix} \quad A_6 = g\begin{pmatrix} 0 & \phi \\ \phi^+ & 0 \end{pmatrix} \tag{4}$$

and the notation is as in Taylor's book (4). This would correspond to a special case of an SU(3) theory if $\lambda = 2$, but we leave it free. Then under a gauge transformation of the form $\begin{pmatrix} U & 0 \\ 0 & 1 \end{pmatrix}$ $U \in SU(2)$ and U independent of x_5, x_6, A_m transforms inhomogeneosly, while ϕ transforms as a doublet $\phi' = U^+\phi$ and under a U(1) transformation of the form

$$\begin{pmatrix} e^{-\frac{i\theta}{\sqrt{2+\lambda^2}}} I & \\ & e^{-\frac{i\lambda\theta}{\sqrt{2+\lambda^2}}} \end{pmatrix} \tag{5}$$

B_μ transforms as $B'_\mu = B_\mu + \frac{g'^{-1}}{\sqrt{2+\lambda^2}} \partial_\mu\theta$ and $\phi' = e^{-\frac{i(\lambda+1)\theta}{\sqrt{2+\lambda^2}}} \phi$

Now suppose that the components of $F_{\mu\nu}$ are calculated we obtain

$$F_{\mu\nu} = \frac{1}{g}\begin{pmatrix} gF_{\mu\nu}(A) - g'F_{\mu\nu}(B) & 0 \\ 0 & \lambda g'F_{\mu\nu}(B) \end{pmatrix}$$

$$F_{\mu 5} = \begin{pmatrix} 0 & iD_\mu\phi \\ -i(D_\mu\phi)^+ & 0 \end{pmatrix} \quad F_{\mu 6} = \begin{pmatrix} 0 & D_\mu\phi \\ (D_\mu\phi)^+ & 0 \end{pmatrix} \tag{6}$$

$$F_{56} = 2g\begin{pmatrix} \phi\phi^+ & 0 \\ 0 & -\phi^+\phi \end{pmatrix}$$

with $D_\mu\phi = (\partial_\mu + gA_\mu - g'(1+\lambda)B_\mu I)\phi$

The lagrangian

$$-\frac{F^2}{4} = -\frac{1}{4}F_{\mu\nu}(A)^2 - \frac{1}{4}\left(\frac{g'}{g}\right)^2 (2+\lambda^2)\, F_{\mu\nu}{}^2(B) - \frac{1}{2}|D_\mu\phi|^2 - g^2(\phi^+\phi)^2$$

Now the choice of constant g' and g in (3) is such that the first component $gA_{11\mu} - g'B_\mu$ is taken to be the field associated with the Z_o, as in Taylor's book.

Let us assume that the theory is arranged such that the kinetic energy terms for the A_μ and the B_μ field are similarly normalised - if $\lambda = 2$ this means that the theory is SU(3) then

$$\frac{g'}{g} = \frac{1}{\left(1+\frac{\lambda^2}{2}\right)^{\frac{1}{2}}} = \tan\theta_W \tag{7}$$

Then the states which couple to the two components ϕ_1 and ϕ_2 of the Higgs field are

$$\left.\begin{aligned} & g\,A_{11\mu} - g'(1+\lambda)B \\ \text{and} & \\ & g\,A_{33\mu} - g'(1+\lambda)B_\mu = -g\,A_{11\mu} - g'(1+\lambda)B_\mu \end{aligned}\right\} \quad (8)$$

Now if we take the first component of the Higgs field ϕ_1 to have zero vacuum expectation value the combination which acquires a mass is $- g\,A_{11\mu} - g'(1+\lambda)B_\mu$. This is orthogonal to the Z_o if $g'^2(1+\lambda) = g^2$

(9)

and identical to the Z_o if $(1+\lambda) = -1$

compatibility with (7) gives in the first case $\lambda = 2$ i.e. SU(3), in the second case $\lambda = -2$ and a value of the Weinberg angle of 30° (5).

Note that the theory so produced is not quite the W-S theory - there is no term quadratic in the mass and there is an extra prediction, namely that the coupling constant of the ϕ^4 term is equal to g^2 but there is no symmetry breaking term. We could try to introduce a term by including diagonal components in the ansatz for A_6 i.e. $A_6 = \begin{pmatrix} -mI & \phi \\ \phi^+ & m \end{pmatrix}$ but since the square of a hermitian quantity $F_{\mu\nu}$ has a trace which is the sum of strictly positive quantities there is no chance of obtaining a relative negative sign of the quadratic compared with the quartic term and hence no symmetry breaking. One method of introducing a quadratic term in a natural way has recently been proposed by Manton (6) following up an observation of Witten (7), that the axially symmetric instanton solutions of SU(2) Yang Mills could be expressed in an ansatz equivalent to a two dimensional theory in a space of constant curvature, incorporating a Higgs field. The idea is to take a metric for the six dimensional space which is not flat in dimensions 5 and 6. Specifically he takes

$$g_{mn} = \mathrm{diag}(1, -1, -1, -1, -\frac{1}{R^2}, -\frac{1}{R^2}\sin^{-2}\theta) \qquad (10)$$

where the additional components of A_i are taken in the polar and azimuthal directions. An alternative formalism is to use a seven dimensional Minkowski space, but restrict the components x_5, x_6, x_7 to lie on the sphere S^2 given by

$$x_5^2 + x_6^2 + x_7^2 = R^2 \qquad (11)$$

Then in terms of the vectors

$$h_1 = A_5 \sin\phi - A_6 \cos\phi$$

$$h_2 = A_5 \cos\phi - A_6 \sin\phi$$

$$h_3 = A_7$$

$$A_\theta = - h_1(x) \qquad (12)$$

$$A_\phi = - h_2(x)\cos\theta + h_3(x)\sin\theta \qquad (13)$$

This idea of spontaneous compactification of the additional dimensions is related to the work of Scherk and Schwarz (8) on symmetry breaking in supergravity theories. R is the parameter which will give the mass. The requirement that the vector fields should be spherically symmetric in the extra dimensions amount to choosing h_1 and h_2 dependent only upon (x), and h_3 as a constant element of the Cartan subalgebra of the gauge group G.

$$[h_3, h_1(x)] = - h_2(x)$$

$$[h_3, h_2(x)] = h_1(x) \qquad (14)$$

$$[h_3\ A_\mu(x)] = 0$$

i.e. $A_\mu(x)$ lies in the little group of h_3

The lagrangian for the theory is

$$\mathcal{L} = \frac{1}{g^2} \int d^6x\ (-\det g)^{\frac{1}{2}} \quad F_{mn}\ F_{st}\ \ g^{mn}\ g^{st} \qquad (15)$$

which reduces on integration over the additional variables to

$$\frac{4\ R^2}{g^2} \int d^4x \quad \{ F^2_{\mu\nu} - \frac{2}{R^2} |(D_\mu h_i)|^2 + \frac{1}{R^4} \{\varepsilon_{ijk}\ h_i + [h_j, h_k], \ \varepsilon_{\ell jk}\ h_\ell + [h_j, h_k]\}\}$$

where $D_\mu h_i = \partial_\mu h_i - [A_\mu, h_i]$ (16)

This is the first advantage of this formalism over the simpler situation discussed earlier (5) - the integration over the extra dimensions gives a finite factor, in (16) since the domain of integration is compact. The second advantage is that a quadratic mass term is generated which is negative with respect to the quartic term and the non constant part of the Higgs potential is essentially

$$\nu(\phi) = - \frac{1}{R^2}\ \phi^+\phi + \alpha g^2 (\phi^+\phi)^2 \qquad (17)$$

where α is a factor dependent upon the gauge group: $\alpha = \frac{1}{2}$ for $G = SU(3)$. Manton determines the Weinberg angle by the requirement that the Higgs field is of hypercharge 1. This then fixes the Weinberg angle as 60° for SU(3) for him. The apparent contradiction with the result of equation (9) where an angle of 30° was obtained is resolved when one asks: what is the charge matrix in Manton's theory?

In the theory presented in reference (5) the charge matrix associated with the fundamental representation is

$$\begin{pmatrix} 0 & & \\ & -1 & \\ & & \frac{\lambda}{2} \end{pmatrix} \qquad (18)$$

whereas in Manton's theory it is

$$\begin{pmatrix} \frac{2}{3} & & \\ & -\frac{1}{3} & \\ & & -\frac{1}{3} \end{pmatrix} \qquad (19)$$

Now if the Weinberg angle is calculated through the formula

$$\sin^2 \theta_W = \frac{T\ \ T_3{}^2}{T\ \ Q^2} \qquad (20)$$

Where T is the third component of weak isospin, then (18) gives $\theta = 30^o$ for $|\lambda| = 2$, while $\theta = 60^o$ for (19). Note however that if the Higgs field with non zero asymptotic limit sits in the (2,3) and (3,2) position in the matrix, then for (18) it only commutes with the charge matrix, i.e. decouples from the photon if $\lambda = -2$, whereas for (19) it already decouples. Thus the two theories require a different interpretation of the basic spinors to which they couple. In Manton's case, the charges would be quark charges, but unfortunately these cannot be associated with the conventional quark multiplets in weak interactions. In the case with $\lambda = -2$ the charges are correct for an interpretation in terms of a $(\nu_L\ e_L)$ doubled and an e_k singlet. This leads to the suggestion that the algebra with $\lambda = -2$ should have a representation in which right and left handed particles belong together. It was noticed independently by Y. Ne'eman (9) and the author that the standard SU(3) matrices $\lambda_4, \lambda_5, \lambda_6, \lambda_7$ close under anticommutation on $\lambda_1, \lambda_2, \lambda_3$ and $\hat{\lambda}_8 = \mathrm{diag}(1,1,2)$ i.e. the simple Lie superalgebra SU(2/1).

Furthermore there exists a four dimensional representation of SU(2/1) with the correct quark charges (10), which also gives rise to the same Weinberg angle (11) provided the quarks are similarly minimally coupled. However there is a serious problem with this idea - the representations always consist of particles of both chiralities i.e. the 3 dimensional representation contains $(\nu_L\ e_L, \tilde{e}_R)$ where $\tilde{e}_R$ is a righthanded negatively charged singlet - which cannot however be identified with the righthanded electron if the other two are to be the electron, neutrino lefthanded doublet because the supergroup implies that the particles associated with different gradings, differ in statistics. In a similar fashion the gauge bosons and the Higgs fields have opposite statistics. This means that if the model is to make any sense the number of particles must be doubled, and some interpretation found for all the ghosts in the theory. There is an ambitious attempt by Thierry Mieg (11, 12) to interpret the unwanted fields in terms of Becchi, Rouet and Stora (13) transformations, and Ne'eman and Thierry Mieg have suggested that the value of the Weinberg angle remains unrenormalised in their theory - the counting of graphs is different from W-S as the ϕ^4 term has a coupling proportional to g^2. However any theory using a supergroup as internal symmetry has the problem of the construction of a positive definite hamiltonian - this has not been satisfactorily solved.

Two other developments are worthy of mention. One scheme has been advocated where the extra dimensions are taken as anticommuting, so that the group manifold becomes now a non compact version of an orthosymplectic supergroup Osp(2,1,2) (14). The potential advantage of this is that the action density remains conformal, and the marriage of a supergroup for space time with one for internal symmetries might avoid problems of statistics - this hope has not yet been successfully realised (15) and perhaps the reason for this is the absence of a theory of differential forms over such manifolds, consistent with the usual formulation of integration. The second point is the absence of a unification scheme comparable with SU(5) - all candidates of the form SU(m/n) have antisymmetric representations of dimension

$$\frac{(n+m)^2}{2} - \frac{(n-m)}{2} \quad \text{and symmetric ones of the form} \quad \frac{(n+m)^2}{2} + \frac{(n-m)}{2}$$

thus the 16 of SU(5/1), as opposed to the conventional $5 + \overline{10}$ of SU(5) contains one particle too many, a massless righthanded neutrino, and also a second photon so must be ruled out. Thus if there is any unified generalisation of the grading idea it must be more subtle than this.

In summary - the attempts to gain a deeper understanding of the Higgs mechanism in terms of group theory have led to a partial completion of the jigsaw, but there are still several pieces missing.

References

1 S.Weinberg, Phys.Rev.Lett. 19 1264 (1979)

A.Salam, Proc. 8th Nobel Symposium Ed. Svartholm (367) Almquist & Wiksell, Stockholm 1968

2 S.W.Hawking, D.N.Page, C.N.Pope, Physics Letters 86B 175 (1978)

3 G. 'tHooft, Nuclear Physics B79 176 (1974)

A.M.Polyakov, JETP Letters 20 194 (1974)

4 J.C.Taylor, Gauge Theories of Weak Interactions, C.U.P. (1976)

5 D.B.Fairlie, Journal of Physics G5 L55 (1979)
Physics Letters B83 979 (1979)

6 N.Manton, LPTENS Preprint 79/12 (1979)

7 E.Witten, Phys. Rev. Lett. 38 121 (1976)

8 J.Scherk and J.H.Schwarz, Nucl. Phys. B153 61 (1979)

9 Y.Ne'eman, Physics Letters 81B 190 (1979)

10 V.Rittenberg, Group Theoretical Methods in Physics, Proc. VI Int. Conf. (Tubingen 1977) P.Kramer and A.Rieches Ed. Springer Verlag.

11 Y.Ne'eman and J.Thierry-Mieg, Tel Aviv Preprint TAUP 727-79 (1979)

12 J.Thierry-Mieg, Thesis, Universite de Meudon, Unpublished (1978)

13 C.Becchi, A.Rouet and R.Stora, Comm. Math. Phys. 42 127 (1975)

14 P.Dondi and P.Jarvis, Phys. Lett. B84 75 (1979)

15 R.Ecclestone, A criticism of supersymmetric Weinberg Salam Models, Southampton preprint 79/15 (1979)

TOPOLOGICAL AND PHENOMENOLOGICAL
ASPECTS OF GRAVITY AND SUPERGRAVITY

Peter G. O. Freund
Enrico Fermi Institute
and Department of Physics
University of Chicago
Chicago, Il. 60637, USA

1. Introduction

It is now generally believed that gravitational, weak, electromagnetic and strong interactions are all to be described by gauge theories. Gauging is essentially a geometrical process and the basic quantities it involves (potentials, field-strengths, etc..) have clear geometrical origin (connections, curvature, etc..) It should, then, not have been as surprising as it actually was when it turned out that topology plays a central rôle in the formulation and solution of crucial physical problems raised by gauge theory. I will review here some of these topological developments as they appear in the theory of gravitation. The two standard topological problems concern solitons and instantons. Gravitational solitons (gravitational magnetic monopoles?) have been explored very little and I will not discuss them. Gravitational instantons have been extensively studied although no clearcut physical uses for them have emerged as yet. Remarkably, however, these studies have led to more general insights into the structure of gauge theory as a whole and I will particularly stress these aspects.

In the second part I will discuss some phenomenological aspects of maximally extended supergravity in which the topological ideas developed in the first part play a certain role.

2. Gravitational Instantons and Generalized Spin Structures

The instantons of nonabelian gauge theory [1] are non-singular classical solutions of the euclidean pure Yang-Mills equations with nontrivial topology. In the same way one can define gravitational instantons as non-singular classical solutions of the Einstein equations with our without cosmological term with (++++) metric-signature, or, in other words, four-dimensional Einstein-manifolds [2,3]. They have two topological invariants that are expressed as integrals over expressions involving the curvature, viz. the Euler-Poincaré charâcteristic

$$\chi = \frac{1}{128\pi^2} \int \frac{\varepsilon^{\alpha\beta\gamma\delta}}{\sqrt{g}} \frac{\varepsilon^{\mu\nu\rho\sigma}}{\sqrt{g}} R_{\alpha\beta\mu\nu} R_{\gamma\delta\rho\sigma} \sqrt{g}\, d^4x \quad + \text{boundary terms} \tag{1a}$$

and the Hirzebruch index

$$\tau = - \frac{1}{96\pi^2} \int \frac{\varepsilon^{\mu\nu\rho\sigma}}{\sqrt{g}} R^{\alpha}{}_{\beta\mu\nu} R^{\beta}{}_{\alpha\rho\sigma} \sqrt{g}\, d^4x \qquad + \text{boundary terms}; \tag{1b}$$

the boundary terms are, of course, absent for compact manifolds.

That there are two such topological invariants is connected with the fact that gravity is the gauge field of the Lorentz-group and, as such, in the "euclidean" sector has gauge group O(4) = O(3)XO(3). The topological invariants τ and χ are related [2,4] to the chiral ABJ-anomaly and the conformal anomaly respectively. This relation, as in the Yang-Mills case, is due to the Atiyah-Singer theorem [1] which connects the topological invariants to zero modes of elliptic operators. Specifically, the index τ of a compact manifold is connected to the difference $n_R^{1/2} - n_L^{1/2}$ of the numbers of right- and left-handed zero modes of the elliptic spin 1/2 (Dirac) operator on that manifold, or, in other words, to the helicity flip ΔQ^5. By contrast, non-compact instantons can contribute [5] only to the chiral zero modes of the spin 3/2 operator. In detail, for the compact case, we find for the helicity flip ΔQ^5 the expression[2,4]

$$\Delta Q^5 = - \frac{N_M}{8} \tau \qquad (2)$$

with N_M the number of fundamental Majorana (i.e., real) spin 1/2 fields in the lagrangian.

At this point let me mention some classes of gravitational instantons (unlike the Yang-Mills case a unified derivation based on algebraic geometry is not available for gravity at this time). Non-compact instantons are the Schwarzschild and Kerr solutions suitably continued to "euclidean" (++++) signature [3], the infinite multi Taub-NUT [7] and EGUCHI-HANSON [5] families, the PAGE-Taub-NUT solution [7]. Compact instantons are the 4 sphere [2], the complex projective plane $P_2(C)$ [2], the metric product S^2XS^2 [7], the S^2 bundle over S^2 of PAGE [8] and the infinite family of 4-dimensional algebraic submanifolds V_{2n} of $P_3(C)$ considered by BACK, FORGER and FREUND [9] and closely related to the work of YAU [10] (I will return to them in more detail below).

Now let me concentrate on these compact instantons. For the 4-sphere and products of 2-spheres $\tau = 0$. The simplest case with $\tau \neq 0$ is $P_2(C)$ for which $\tau = 1$. But ΔQ^5 should be an even integer, whereas, according to (2), $\Delta Q^5 = - N_M/8$ in this case. In particular, for $N_M = 1$ this yields the nonsensical value $\Delta Q^5 = - 1/8$. The reason is that the anomaly (ΔQ^5) is due to Feynman diagrams involving spinor fields, but these fields cannot live globally on $P_2(C)$. How can this happen?

Given a vierbein we can always locally define spinors on a Riemann 4-manifold. A global definition on the whole manifold is possible only if its second Stiefel-Whitney class w_2 vanishes. This is connected with the properties of closed curves in the orthonormal frame bundle that lie entirely in one fiber and correspond to vierbein rotations of 2π so that spinors change sign. Such curves cannot be contracted to a point while staying in the fiber. But by deforming them through distant regions of the bundle, their contraction to a point can still be achieved if $w_2 \neq 0$. The instructions for spinor signs are then self-contradictory, and no global definition of spinors is possible. The simplest procedure for avoiding this problem is to construct "generalized spin structures" on the manifold. [9,11,12] Essentially, one assigns spinors to multiplets of some gauged internal symmetry and absorbs the undetermined sign in the gauge indeterminancy.

The problem is thus to devise a procedure that will upgrade the orthonormal frame bundle (an SO(4) bundle) to a bundle that allows the global definition of spinors. If $w_2 \neq 0$ then the SO(4) bundle cannot just be upgraded to a Spin(4) $\equiv$ SU(2)XSU(2) bundle and the question is what group to use instead of Spin(4). The simplest possibility [9] turns out to be $\overline{\text{Spin}}(4) = \text{Spin}(4)X_{Z_2}SU(2)$. Indeed, there exists a homomorphism. λ: Spin(4) $\rightarrow$ Spin(4)XSU(2) defined by the identity map to Spin(4) and projection to one of its SU(2) factors as the map to SU(2). We now specify $\overline{\text{Spin}}(4)$ by identifying $(g,h)\sim(-g,-h)$ for $g\varepsilon$Spin(4), $h\varepsilon$SU(2). The groups SO(4) and $\overline{\text{Spin}}(4)$ differ from Spin(4) and Spin(4)XSU(2) each by a Z_2 factor. The just described λ-homomorphism then induces an SO(4) $\rightarrow$ $\overline{\text{Spin}}(4)$ homomorphism that can be used to upgrade the SO(4)-bundle into a $\overline{\text{Spin}}(4)$ bundle and thus to consistently define spinors on any Riemann 4-manifold once one has gauged SU(2). But for this to work there is one more requirement. Any representation Λ of $\overline{\text{Spin}}(4)$ on a vector space can be pulled back to a representation of Spin(4)XSU(2) $\equiv$ $(SU(2))^3$. But the element (-1, -1, -1) of $(SU(2))^3$ must map to the identity in Λ. Label Λ by three spins (j_1, j_2, j_3). This, then, requires $e^{i\pi(2j_1+2j_2+2j_3)} = 1$ or $j_1 + j_2 + j_3$ = integer. But the ordinary spin-statistics connection tells us that $j_1 + j_2$ = half-odd integer (integer) for fermions (bosons). Thus we find that the "internal spin" j_3 must be half-odd integer (integer) for fermions (bosons) and thus establish an "internal spin" - statistics connection.

This construction can be readily generalized to other nonabelian Lie groups [9,13] that contain a Z_2 in their center. For any such group G, replace Spin(4) in the above argument by $Spin_G(4) \equiv Spin(4) X_{Z_2} G$ (the construction above had G = SU(2)). G = SU(2)XSU(2) gives an amusing phenomenology [9] if identified as the $SU(2)_L XSU(2)_R$ of electroweak gauge theory. We shall, rather, go on directly to "grand" unification theories. Such theories as a rule, postulate some "very" large simple group that encompasses $SU(3)_{color} X(SU(2)XU(1))_{electroweak}$. Typically, SU(N) and SO(N) groups have been considered. Since the "grand" unification group is the maximal gauged internal symmetry group, it is the only candidate of G in $Spin_G(4)$. Now for SU(N) groups, only those with even N have a Z_2 in their center (those with N = odd have Z_N but no Z_2 in their center). This rules out the famous SU(5) of GEORGI and GLASHOW [14]. (For SO(N) groups rather than a selection in N, we are instructed to take the twofold covering Spin(N) which is possible for all N > 3). We thus have a theoretical criterion for the selection of a grand unified gauge group. Weak as it is, it manages to rule out the much discussed SU(5)-theory. We shall see more of this later for extended supergravity.

Following this discussion, we can now show how the anomaly problem is taken care of. On a manifold without a usual spin structure $N_M^{\tau}/8$ is not an integer. But when we define the generalized spin structure $Spin_G(4)$ there is an extra contribution to ΔQ^5 corresponding to the Pontryagin number of the gauge field of the group G that defines $Spin_G(4)$. Now this is not integer either but the sum of the gravitational and G-gauge field contributions is an even integer as required. There is a more general aspect of all this. With a generalized spin structure the quantum theory can be defined (in the "euclidean" region) by a path integral over the gravitational field covering all Riemann manifolds. Without the generalized spin structure, i.e., without the extra gauged symmetry G one would have to restrict the path integral to spin manifolds (this is natural in certain supergravity theories).

Now back to the instanton list. The compact manifolds V_{2n} mentioned earlier are all spin manifolds. Let us examine them a little closer with the anomaly in mind. They are obtained by looking for algebraic submanifolds of $P_3(C)$. $P_3(C)$ has four homogeneous complex coordinates z_1, z_2, z_3, z_4 such that $\Sigma |z_i|^2 = a^2$ and points with coordinates z_i and $e^{i\alpha}z_i$ are to be identified. Now let $F_m(z_i)$ be a homogeneous polynomial in the z_i of degree m, such that grad $F_m \neq 0$ for $z \neq 0$. Then $F_m(z_i) = 0$ defines a Kähler submanifold of $P_3(C)$. For even m = 2n the second Stiefel-Whitney class of V_m vanishes so that it is a spin-manifold. Standard topological calculations yield

$$\tau(V_{2n}) = -16 \binom{n+1}{3}$$

so that $\tau(V_2) = 0$, $\tau(V_4) = 16, \ldots$ ΔQ^5 is then always an even integer ($\tau(V_{2n})$ being an integer multiple of 16) and no "paradox" of the $P_2(C)$ type appears. The V_4's are the much discussed K3 surfaces of the mathematicians. YAU's brilliant proof of CALABI's conjecture [10] means that each V_4 manifold admits a Ricci flat metric. Unfortunately, its explicit construction has not been achieved as yet. The V_{2n} manifolds with n > 2 all obey the Hitchin inequalities between χ and τ so that they could admit Einstein metrics with $\Lambda < 0$.

All gravitational instantons known so far are algebraic manifolds as are all Yang-Mills instantons (and all of those are known). What does this mean? Consider a field theory in one-time and zero space dimensions, i.e., a point particle in a potential. Take central potentials; then there always are classical solutions (trajectories) that are algebraic curves, namely, circles. But the counterpart of what is found for gauge theories would be the requirement that all trajectories be algebraic curves. Then all bounded trajectories must close and by Bertrand's theorem the potential is either of the Coulomb or harmonic-oscillator form. But these are the two cases in which there are extra conservation laws (Lenz vector, quadrupole tensor) and which can be solved in closed form at both the classical and quantum levels. Could it be that gauge theories also have unexpected (presumably infinitely many) conservation laws and are exactly soluble?

3. Phenomenology of Supergravity

Ordinary and extended supergravities [15] are the gauge theories of $OSp(N|4)$ supergroups $N=1,\ldots,8$. For $N > 1$ they involve gauged internal symmetries. To see this, let us look more closely at the $OSp(N|4)$ Lie-superalgebras [16]. Consider a graded vector-space V such that any vector $v \varepsilon V$ has a grade $gr(v) \varepsilon Z_2$. Let the dimension of the Bose (grade zero) sector of V be N and that of its Fermi (grade one) sector be 2M (we obviously imply a superselection rule forbidding the linear combination of Bose and Fermi vectors). The grading of V induces an obvious Z_2 grading on the linear transformations of V. These form a Lie superalgebra under the binary bracket operation defined to be a commutator in all cases but one, namely, when both transformations are Fermi, in which case, it is defined as the anticommutator (this bracket obeys a graded Jacobi identity and is grade additive mod 2). Represent the vectors of V as column matrices with N+2M entries. With the metric tensor

$$G = \begin{pmatrix} C & 0 \\ 0 & 1 \end{pmatrix} \begin{matrix} 2M \\ N \end{matrix} \quad (\text{columns: } 2M, N) \qquad 1 = N\mathrm{x}N \text{ unit matrix} \qquad C = \begin{pmatrix} i\sigma_2 & & \\ & i\sigma_2 & \\ & & \ddots \end{pmatrix} \tag{3a}$$

we define the scalar product

$$(v,w) = v^T G w \tag{3b}$$

for $v, w \varepsilon V$. On the linear transformations on V we now impose the "superantisymmetry" condition

$$(v, Tw) + (Tv, w)(-1)^{gr(T)gr(v)} = 0 \ . \tag{4a}$$

In matrix form

$$T = \begin{pmatrix} a & p \\ q & b \end{pmatrix} \begin{matrix} 2M \\ N \end{matrix} \quad (\text{columns: } 2M, N) \tag{4b}$$

so that "superantisymmetry" requires

$$a^T = CaC \qquad b^T = -b \qquad q = -p^T C \tag{4c}$$

the set of all such T generate $OSp(N|2M)$. There are 2MN Fermi generators and $\frac{N(N-1)}{2}$ + +M(2M+1) Bose generators that span $Sp(2M) \mathrm{X} O(N)$ as is clear from (4). The Lie algebra $Sp(4)$ is isomorphic to the de Sitter algebra $O(3,2)$. Thus, $OSp(N|4)$ has, in its Bose sector, the direct sum of the de Sitter algebra and of an internal $O(N)$ algebra. The gauging of $OSp(N|4)$ thus involves, among others, the gauging of the de Sitter algebra i.e., Einstein gravitation with a cosmological term and an $O(N)$ Yang-Mills theory. The supermultiplets of such a theory contain [17] one particle of maximum helicity, say, λ and $\binom{N}{m}$ particles of helicity $\lambda - \frac{m}{2}$ for $m = 1,2,\ldots,$ all the way to minimum helicity $\lambda - \frac{N}{2}$. If all particles in a supermultiplet are to have spins ≤ 2, then both $|\lambda|$ and $|\lambda - N/2|$ have to be ≤ 2 so that $N \leq 8$. In particular, for $N=8$ one must have $\lambda = 2$. We have

helicity	2	3/2	1	1/2	0	-1/2	-1	-3/2	-2
O(8) multiplet	1	8	28	56	35+35	56	28	8	1

Phenomenologically, this is rather unsatisfactory as O(8) does not even contain $SU(3)_{color} X (SU(2) X U(1))_{electroweak}$. Even if one is willing to "compromise" away the SU(2) part of the electroweak gauge group [18], there are too few color triplets and singlets in the spin 1/2 sector to accomodate all the known quarks and leptons. In short, then, this phenomenology is not particularly successful.

Recent work by CREMMER, JULIA and SCHERK [19] and by NAHM [20] has revealed a hidden nonlinearly realized SU(8) gauge invariance in N=8 supergravity. This means that gauge invariance is achieved not with the vector fields of the theory - of which, anyway, there wouldn't be sufficiently many - but, rather, through nonlinear combinations of the scalar fields (like in CP^{n-1} σ-models). Yet, quantum theory may produce the vector gauge particles of SU(8) as bound states. A similar phenomenon does happen in 2-dimensional CP^{n-1} σ-models. This raises the complex dynamical question of bound states in the theory. Yet, hopefully, once one includes the bound states, one will find all the 63 gauge bosons of SU(8). Among other things, supersymmetry then requires also some additional spin 1/2 fermions in the spectrum. This is a welcome addition and has led to a phenomenological reappraisal of N=8 supergravity by CURTRIGHT and myself [21].

To obtain the 63 gauge bosons we note the SU(8) → SO(8) branching rule 63 → 28 + 35 where 28 and 35 are, respectively, the antisymmetric and symmetric tensor representations of SO(8). SO(8) happens to have two more inequivalent bispinorial 35-dimensional representations that are interchanged with the symmetric tensorial one via Cartan's "triality" transformations. We, therefore, do not distinguish between these three à priori inequivalent 35-multiplets. From this observation it then follows that the supermultiplets that yield 28 and 35 at the spin one level have maximal helicity $\lambda = 2$ and $\lambda = 3$ respectively (the spin 3 and 5/2 particles of the $\lambda = 3$ supermultiplet are harmless as they do not correspond to fields in the basic lagrangian but to bound states). Of these, the $\lambda = 2$ supermultiplet is CPT self-adjoint, while the $\lambda = 3$ supermultiplet has a $\lambda = 1$ CPT partner. The $\lambda = 2$ supermultiplet gives a 56-dimensional irreducible representation of both SU(8) and SO(8) at the spin 1/2 level. The $\lambda = 3$ multiplet also gives a 56 but also an 8. There may be many more bound state supermultiplets and, correspondingly, the spin 1/2 spectrum may be very rich. Whichever way we look at it, however, we cannot fail to notice at least two 56 spin 1/2 multiplets. With those, and a judicious embedding of $SU(3)_{color} X (SU(2) X U(1))_{electroweak}$ into SU(8) we can construct a rather satisfactory picture of quarks and leptons.

Before going into the details, let me point out that with two Majorana 56 multiplets - or equivalently, a left-handed and a right-handed one - we can construct a vector-like model, thus cancelling undesirable ABJ-anomalies. Moreover, as we shall find weak isosinglets along with isodoublets, we shall also avoid the troublesome parity conserving neutral currents of earlier vector-like models. Strictly from the point of view of anomaly cancellation a "chiral" model with spin 1/2 particles in the $56+\overline{28}+\overline{8}$ representations of SU(8) is also tenable. Remarkably, it contravenes to the "internal-spin" - statistics connection of section 2. The Young tableaux of 56 and $\overline{8}$ have odd number of boxes, while that of $\overline{28}$ has an even number of boxes. The internal spin statistics connection forbids the latter. The connection is automatic in supermultiplets with integer maximal helicity (in the vector-like model we only had $\lambda = 1,2,3$) but is violated in those with half-odd-integer λ such as the $\lambda = 5/2$, 3/2 that contain the $\overline{28}$ of spin 1/2. Because of this, I will not further discuss the "chiral" model here (it is not particularly natural as far as supergravity is concerned either). It is also relevant to note that the 28 of spin 1 being in a $\lambda = 2$ supermultiplet, one may be tempted to say that this is the fundamental input multiplet of the whole theory. A priori, this seems unlikely since the absence of a cosmological term renders the 28 vector particles of this supermultiplet into gauge bosons of central charges rather than of O(8). In quantum theory this may change. Yet at the spin 1/2 level we still get a 56-multiplet so, whatever its origin, the phenomenology is unaffected. Supergravity, then, strongly suggests a spin 1/2 spectrum that starts with two 56 multiplets, and possible further (many?) other multiplets.

Now to the details of the vector-like model with a 56_L and a 56_R. Unlike previous SO(8) supersymmetric phenomenology, with SU(8) one easily finds an $SU(3)_{color}X(SU(2)XU(1))_{electroweak}$ subgroup with symmetry to spare. To explore the quark-lepton spectrum, break SU(8) down to SU(3)XSU(2)XU(1) through the intermediate step $SU(3)_{color}XSU(3)_{shape}X(SU(2)XU(1))_{electroweak}$. Here $SU(3)_{shape}$ is a "horizontal" symmetry [22] that shuffles the quarks $u \leftrightarrow c \leftrightarrow t$, $d \leftrightarrow s \leftrightarrow b$ and the leptons $e \leftrightarrow \mu \leftrightarrow \tau$; $\nu_e \leftrightarrow \nu_\mu \leftrightarrow \nu_\tau$. To appreciate this, first consider the fundamental 8 multiplet of SU(8). We postulate that each non-Abelian attribute - color, shape, and weak isospin - is exclusively carried by states in this fundamental 8. This means: the carrier triplet of color (chromons) should have neither shape nor weak isospin, the triplet carrier of shape (morphons) should have neither color nor weak isospin, and the doublet carrier of weak isospin (asthenons) should have no color and no shape. Designate multiplets having color (shape) multiplicity a_{color} (a_{shape}) and weak isospin I with the notation (a_{color}, a_{shape}, I). Then the branching rule is

$$8 \rightarrow (3, 1, 0) + (1, 3, 0) + (1, 1, 1/2).$$

Exclusivity suggests interpreting the eight members of the fundamental octet as basic constituents of quarks and leptons. This is similar though not identical to earlier subquark models [23]. The quarks and leptons in the 56 can then be viewed as built out of three such basic constituents. In particular, the (3, 3, 1/2) and (1, $\overline{3}$, 1/2) multiplets that appear in the 56 must have electric charge values of (2/3, - 1/3) and (1, 0) to be identified with quarks and antileptons, respectively. This identification and the requirement that the electric charge operator be an SU(8) generator (i.e., traceless) fix the electric charges of the chromons, morphons, and the asthenon doublet to be -1/3, 0 and (1, 0) respectively. From this, we can now determine the color, shape, weak isospin and electric charge assignments of any SU(8) multiplet.

The contents of the left-handed and right-handed 56 are displayed in the Table on the next page.

All color singlets are integrally charged and thus correspond to antileptons (Q=1,0) and leptons (Q=0,-1) while all colored states are fractionally charged and appear in either color triplets (quarks) ar anti-triplets (antiquarks). No "exotic" $|Q| > 1$ charges, and (unlike in previous work [18]) no color sextets nor octets appear. Note there are three new neutrino-like particles, the N-leptons, which must be massless. Also note that there are right-handed doublet currents involving the u, c, and t quarks, given the assignments in column 1. Regarding these currents, present experiments seem to require [24] $m_{x,y,z} \gtrsim 8$ GeV/c². However, replacing $\binom{u\ c\ t}{x\ y\ z}_R \rightarrow \binom{f\ g\ h}{x\ y\ z}_R$ and $(\overline{f}\ \overline{g}\ \overline{h})_L \rightarrow (\overline{u}\ \overline{c}\ \overline{t})_L$ avoids such right-handed currents and mass constraints. The experimentally observed threefold replication ("sequential nature") of both the known leptons and quarks is simply accounted for in the Table by assigning these leptons and quarks to triplets of broken ($m_d \neq m_s \neq m_c$) $SU(3)_{shape}$ symmetry. For simplicity, we have suppressed all Cabibbo-like mixing angles in the Table.

A vector-like SU(8) theory with a 56 of spin 1/2 fermions and electric charge assignments as in the Table was considered by GELL-MANN, RAMOND, and SLANSKY [25] in their classification of infinitely many grand unification schemes obeying certain minimality requirements. They did not anticipate a connection with N=8 supergravity and did not discuss $SU(3)_{shape}$. However, they did discuss baryon number conservation for the vector-like model and we refer to their paper for details. Here, it suffices to say that the model is compatible with the existing lower bound on the proton lifetime.

The details of the Higgs-Englert-Brout mechanism for supergravity with a broken local SU(8) symmetry should be quite complicated due to the composite nature of some

Table Quarks and Leptons in the SU(8) Models

Particles in the Vector-like model	color	shape	weak isospin	electric charge
$\begin{pmatrix} u & c & t \\ d & s & b \end{pmatrix}_L \begin{pmatrix} u & c & t \\ x & y & z \end{pmatrix}_R$	3	3	1/2	$\begin{pmatrix} 2/3 \\ -1/3 \end{pmatrix}$
$(x\ y\ z)_L\ (d\ s\ b)_R$	3	$\overline{3}$	0	-1/3
a_L a_R	3	1	0	2/3
$(\overline{f}\ \overline{g}\ \overline{h})_L\ (\overline{f}\ \overline{g}\ \overline{h})_R$	$\overline{3}$	3	0	-2/3
$\begin{pmatrix} \overline{v} \\ \overline{w} \end{pmatrix}_L \begin{pmatrix} \overline{v} \\ \overline{w} \end{pmatrix}_R$	$\overline{3}$	1	1/2	$\begin{pmatrix} 1/3 \\ -2/3 \end{pmatrix}$
$\begin{pmatrix} \overline{E} & \overline{M} & \overline{T} \\ \overline{N}_E & \overline{N}_M & \overline{N}_T \end{pmatrix}_L \begin{pmatrix} \overline{e} & \overline{\mu} & \overline{\tau} \\ \overline{\nu} & \overline{\nu}_\mu & \overline{\nu}_\tau \end{pmatrix}_R$	1	$\overline{3}$	1/2	$\begin{pmatrix} 1 \\ 0 \end{pmatrix}$
$(\overline{e}\ \overline{\mu}\ \overline{\tau})_L\ (\overline{E}\ \overline{M}\ \overline{T})_R$	1	3	0	1
ζ_L ζ_R	1	1	0	-1
η_L η_R	1	1	0	0

of the bosons. On a phenomenological level, the λ = 2 and 3, N = 8 supermultiplets suggest using a pair of opposite parity, spin zero adjoint representations of SU(8).

We have described the simplest viable model of quarks and leptons suggested by N = 8 supergravity. However, there may be more spin 1/2 states for that theory depending on the total wealth of its bound state spectrum. The model described here is offered as a lower bound to the wealth of that spectrum. Also, since gravitation is present in that theory, we are presumably dealing with a model where the mass region below the Planck mass consists of a small population of relatively low mass, fundamental states, followed by a "desert" populated by lesser composite states.

In four dimensions, N = 8 supergravity looks neigher simple nor particularly beautiful. Its only aesthetic interpretation is that of ordinary (i.e., unextended N = 1) supergravity in eleven dimensions, reduced to four [19,20]. The original eleven dimensional space is singled out by the nonexistence of supergravity in higher dimensions [20]. This raises the question of why precisely seven of the ten space-like dimensions "curl up", and is, perhaps, the first really meaningful formulation of the age-old question "why do we experience one time-like and three space-like dimensions?" Also note that spin 1/2 and 0 matter fields in four dimensions are remnants of gauge fields (i.e., force fields) in eleven dimensions. The concepts of matter and force are then different facets of one and the same phenomenon in N = 8 supergravity.

Ideas involving supergravity phenomenology, SU(8) grand unification as one possibility, and horizontal symmetries have all been separately considered before [18, , 22]. The key new point made is that these ideas could fit together in a natural and harmonious manner.

References

1. See, e.g., S. Coleman, Harvard preprint HUTP/A004.

2. T. Eguchi and P.G.O. Freund, Phys. Rev. Lett. 37, 1251 (1976).

3. S.W. Hawking, Phys. Lett. 60A, 84 (1977); see also, A.A. Belavin and D.E. Burlankov, Phys. Lett. 58B, 7 (1976).

4. R. Delbourgo and A. Salam, Phys. Lett. 40B, 3881 (1972); M.J. Duff, Nucl. Phys. B125, 334 (1977; S.M. Christensen and M.J. Duff, Harvard preprint December 1978.

5. S.W. Hawking and C.N. Pope, Nucl. Phys. B146, 381 (1978).

6. T. Eguchi and A. Hanson, Phys. Lett. 74B, 249 (1978); G.W. Gibbons and S.W. Hawking, Phys. Lett. 78B, 430 (1978); N. Hitchin, Proc. Cambridge Phil. Soc. (in press).

7. G.W. Gibbons and S.W. Hawking, Comm. Math. Phys. 66, 291 (1979).

8. D.N. Page, Phys. Lett. 79B, 235 (1978).

9. A. Back, P.G.O. Freund and M. Forger, Phys. Lett. 77B, 181 (1978).

10. S.T. Yau, Proc. Nat. Acad. Sci. USA 74, 1798 (1977).

11. S.W. Hawking and C.N. Pope, Phys. Lett. 73B, 42 (1978).

12. For compact Riemann 4-manifolds $Spin^c(4) = Spin(4)X_{Z_2}U(1)$ structures can always be defined: F. Hirzebruch and H. Hopf, Math. Ann. 136, 156 (1958); for $P_2(C)$ the relevant U(1) gauge field is given in A. Trautman Int. J. Theor. Phys. 16, 561 (1977).

13. M. Forger and M. Hess, Comm. Math. Phys. 64, 269 (1979); S.P. Avis and C.J. Isham, preprint ICTP/78-89/21; S. Hawking and M. Rocek, private communication.

14. M. Georgi and S. L. Glashow, Phys. Rev. Lett. 32, 438 (1974).

15. D.Z. Freedman, P. van Nieuwenhuizen and S. Ferrara, Phys. Rev. D13, 3214 (1976); S. Deser and B. Zumino, Phys. Lett. 62B, 325 (1976).

16. P.G.O. Freund and I. Kaplansky, J. Math. Phys. 17, 228 (1976).

17. A. Salam and J. Strathdee, Nucl. Phys. B80, 499 (1974); M. Gell-Mann and Y. Ne'eman, unpublished.

18. M. Gell-Mann, Talk at the 1977 Washington Meeting of the American Physical Society.

19. E. Cremmer, B. Julia and J. Scherk, Phys. Lett. 76B, 409 (1978); E. Cremmer and B. Julia preprint LPTENS 79/6.

20. W. Nahm, Nucl. Phys. B135, 149 (1978).

21. T.L. Curtright and P.G.O. Freund, preprint EFI 79/25.

22. F. Wilczek and A. Zee, Phys. Rev. Lett. 42, 421 (1979).

23. O.W. Greenberg, Phys. Rev. Lett. 35, 1120 (1975); J.C. Pati and A. Salam, Phys. Lett. 59B, 265 (1975).

24. H. Fritzsch, Proc. 19th Int. Conf. High Energy Physics, S. Homma, M. Kawaguchi, M. Miyazawa editors, Phys. Soc. of Japan 1979, p. 593.

25. M. Gell-Mann, P. Ramond and R. Slanky, Rev. Mod. Phys. 50, 721 (1978).

BIFURCATION AND STABILITY IN YANG-MILLS THEORY WITH SOURCES

R. Jackiw
Center for Theoretical Physics
Laboratory for Nuclear Science and Department of Physics
Massachusetts Institute of Technology
Cambridge, Massachusetts 02139

1. Introduction

In my lecture I shall discuss some recent work on solutions to classical Yang-Mills theory. The investigations that I shall summarize study the field equations with static, external [non-dynamical] sources. The physical, quantum-mechanical significance of such solutions has not thus far been as profound as that of solitons, where sources are dynamical i.e. monopoles with Higgs-field sources; nor as that of instantons, where sources are absent but the equations are continued to imaginary time. Nonetheless the new results are interesting in their differences from the Abelian counterpart and should suggest intuition about the physical content of non-Abelian gauge-quantum field theory. Moreover, structurally the equations are sufficiently intricate to provide a most interesting example of analysis in mathematical physics.

There is now available a variety of solutions for review; but only recently did a pattern emerge which allows for a comprehensive description. A summary of this is presented in Section 2. Section 3 is devoted to an account of stability properties. I begin by recalling the theory of stability -- a subject widely studied by physicists in former times, but now, in its general form, largely forgotten. The general theory does not rely on minima of the energy and is found to be applicable to the Yang-Mills model. It comes as no surprise that the non-Abelian structure lets the gauge theory share with a top the phenomenon of stable configurations which do not minimize the energy. The Section concludes with an assessment of the stability of various solutions. Finally a list of open questions and problems for further research comprise the concluding Section 4.

2. Solutions to Yang-Mills Theory with Sources

2.1 Preliminaries

The field equations with which we are concerned are

$$\mathcal{D}_\mu F^{\mu\nu}=\delta^{\nu o}\rho \tag{2.1a}$$

$$F^{\mu\nu}=\partial^\mu A^\nu-\partial^\nu A^\mu+[A^\mu,A^\nu] \tag{2.1b}$$

$$\mathcal{D}_\mu=\partial_\mu+[A_\mu, \tag{2.1c}$$

We study the SU(2) theory with coupling strength scaled to unity and use interchangeably component notation and anti-Hermitian matrix notation; e.g. ρ_a, $a=1,2,3$; $\rho=\rho_a\sigma^a/2i$, σ^a=Pauli matrices. The source ρ is taken to be a given, time-independent function, $\partial_t\rho=0$. Eq. (2.1a) carries with it an integrability condition; the right-hand side must be covariantly conserved. In the present circumstance that requirement reduces to

$$[A^o,\rho] = 0 \tag{2.2}$$

The energy of the system is given by a positive, gauge-invariant formula

$$\mathcal{E} =\tfrac{1}{2}\int d\vec{r}\ \{\vec{E}_a^2 + \vec{B}_a^2\}$$
$$E_a^i = F_a^{io},\quad B_a^i = -\tfrac{1}{2}\varepsilon^{ijk}F_a^{jk} \tag{2.3}$$

The class of solutions which I shall here be describing is delimited by the requirement of finite energy. This means that sources must also be well-behaved; a condition which will not be spelled out in detail, beyond noting that point sources are excluded; ρ is an extended function.

When it comes to discuss stability, we shall want a Hamiltonian formulation for the field equations (2.1); since they are locally gauge invariant this is not straightforward. In order to overcome the familiar difficulty, we do not pick a gauge; rather we take the variations used for obtaining the equations of motion to be constrained by Gauss' law, which is the $\nu=0$ component of (2.1a). Specifically we take the Hamiltonian to coincide with $\mathcal{E}$, viewed as a functional of independent variables $\vec{E}$ and $\vec{A}$, while $\vec{B}$ is constructed in the usual way from $\vec{A}$.

$$\vec{B}_a =\vec{\nabla}\times\vec{A}_a - \tfrac{1}{2}\varepsilon_{abc}\vec{A}_b\times\vec{A}_c \tag{2.4}$$

$-\vec{E}$ is identified with the canonical momentum conjugate to $\vec{A}$, and the constraint of Gauss' law is imposed with the help of a Lagrange multiplier, here called A_a^o. Hence unrestricted variation can be performed on

$$\bar{\mathcal{E}} = \mathcal{E} - \int d\vec{r} A_a^0 (\vec{\nabla}\cdot\vec{E}_a - \varepsilon_{abc}\vec{A}_b\cdot\vec{E}_c - \rho_a) \tag{2.5}$$

In this way the Yang-Mills equations are obtained.

$$0 = -\frac{\delta\bar{\mathcal{E}}}{\delta A_a^0} = \vec{\nabla}\cdot\vec{E}_a - \varepsilon_{abc}\vec{A}_b\cdot\vec{E}_c - \rho_a \quad \text{(Gauss' law constraint)} \tag{2.6a}$$

$$\partial_t\vec{E}_a = \frac{\delta\bar{\mathcal{E}}}{\delta\vec{A}_a} = \vec{\nabla}\times\vec{B}_a - \varepsilon_{abc}\vec{A}_b\times\vec{B}_c - \varepsilon_{abc}A_b^0\vec{E}_c \quad \text{(Ampère's law)} \tag{2.6b}$$

$$\partial_t\vec{A}_a = -\frac{\delta\bar{\mathcal{E}}}{\delta\vec{E}_a} = -\vec{E}_a - \vec{\nabla}A_a^0 + \varepsilon_{abc}\vec{A}_b A_c^0 \quad \text{(Definition of } \vec{E}_a) \tag{2.6c}$$

Note that static solutions [all time-derivatives vanish] are critical points of the energy, subject to the constraint of Gauss' law [1,2].

Presentation of solutions is complicated by the gauge covariance of (2.1): if A^μ solves the equations with source ρ, then the equations with a gauge-rotated source ρ'

$$\rho' = U^{-1}\rho U \tag{2.7a}$$

are solved by gauge transforming the previous.

$$A'^\mu = U^{-1}A^\mu U + U^{-1}\partial^\mu U \tag{2.7b}$$

[Here U is an SU(2) matrix.] Two solutions related as above describe the same physical situation and we shall view them as the same solution but presented in different "gauge frames". Frequently we shall speak of an "Abelian gauge frame" -- one in which the source points in the third direction.

$$\rho_a = \delta_{a3}q \tag{2.8}$$

Of course results for gauge invariant quantities, like the energy, are frame independent.

In addition to the above gauge covariance, there is present also a gauge invariance with respect to gauge transformations which leave the source unchanged. From (2.2), we see that gauge transformations with U constructed from A^0 are of this type. Thus it is always possible to pass to the temporal gauge where A^0 vanishes, without changing the gauge frame.

Solutions naturally fall into two classes: those that exist for arbitrary sources and those that require a critical, finite source strength. We list these in turn.

2.2 Arbitrary Sources

Four different types of solutions will be discussed in this sub-Section, two static, two time-dependent. The latter provide a well-defined generalization of the former.

The most obvious Yang-Mills solution is the static Coulomb one which is readily presented in the Abelian gauge frame, where it is given by the regular solution to Poisson's equation [3].

$$A_a^o = \delta_{a3}\phi \tag{2.9a}$$

$$\vec{A}_a = 0 \tag{2.9b}$$

$$\phi = -\frac{1}{\nabla^2}q \tag{2.9c}$$

An alternate description, still in the Abelian gauge frame, is gotten by passing to the temporal gauge.

$$A_a^o = 0 \tag{2.10a}$$

$$\vec{A}_a = \delta_{a3}\vec{\nabla}\phi t \tag{2.10b}$$

The energy of this, according to (2.3), is the familiar Coulomb expression.

$$\mathcal{E}_c = \tfrac{1}{2}\int q\,\frac{-1}{\nabla^2}\,q = \frac{1}{8\pi}\int d\vec{r}d\vec{r}'\,\frac{q(\vec{r})q(\vec{r}')}{|\vec{r}-\vec{r}'|} \tag{2.11}$$

Note that in the Abelian frame, the solution vanishes with the source.

The next solution is a time-dependent generalization of the above. It shares with the Coulomb solution the [gauge-invariant] property that the magnetic field vanishes. From Ampère's law it follows that, in the temporal gauge, vanishing $\vec{B}$ implies a static electric field. Eqs. (2.4) and (2.6c) require the electric field to be [gauge equivalent to] a gradient of a scalar [matrix] function Φ, which further must satisfy

$$[\vec{\nabla}\Phi, \vec{\nabla}\Phi] = 0 \tag{2.12}$$

a condition which is easily fulfilled; see Ref. [4]. Thus we have, for Φ's satisfying (2.12),

$$A'^o = 0 \tag{2.13a}$$

$$\vec{A}' = \vec{\nabla}\Phi t \tag{2.13b}$$

[A primed notation is used as reminder that the solution is being presented in a gauge frame other than the Abelian one.] The source which gives rise to such a field is determined by Gauss' law.

$$\nabla^2 \Phi = -\rho' \tag{2.13c}$$

There are as many configurations in the above category as there are functions Φ consistent with (2.12). However, our interest is only in those for which ρ' is gauge equivalent to the Abelian frame formula (2.8); only then are we dealing with solutions to the same problem as the Coulomb one.

$$\rho' = U\rho U^{-1} = U \frac{\sigma^3}{2i} U^{-1} q \tag{2.14}$$

When (2.14) holds, we can express the solution in the Abelian frame, in the temporal gauge.

$$A^0 = 0 \tag{2.15a}$$

$$\vec{A} = -\vec{E}t - U^{-1}\vec{\nabla}U \tag{2.15b}$$

$$\vec{E} = -U^{-1}\vec{\nabla}\Phi U \tag{2.15c}$$

The energy of the above is given by a Coulomb-type formula.

$$\mathcal{E} = \tfrac{1}{2}\int \rho'_a \frac{-1}{\nabla^2} \rho'_a = \frac{1}{8\pi} \int d\vec{r}d\vec{r}' \frac{\rho'_a(\vec{r})\rho'_a(\vec{r}')}{|\vec{r}-\vec{r}'|} \tag{2.16}$$

To recapitulate, the solution for a given source (2.8) is constructed by choosing a gauge function U; computing ρ' from (2.14); Φ from (2.13c); and finally, the potentials from (2.13a) and (2.13b) or (2.15). When (2.12) is met, one has solutions which in general are essentially time-dependent -- a time-translation cannot be compensated by a gauge-transformation. The only member of the family with gauge-artifactual time-dependence is the Coulomb one where U=I. By continuously deforming I to U, one passes continuously from the static Coulomb solution to its time-dependent generalization [4]. In Section 3 it will be demonstrated that the energy (2.16) can be lowered by an arbitrary amount below the Coulombic value $\mathcal{E}_c$.

Our third solution is again static, but it differs from the Coulomb one by the property that in the Abelian frame it does not vanish with the source; rather it becomes a pure gauge.

$$A^\mu \Big|_{\substack{\text{zero}\\ \text{source}}} = U^{-1}\partial^\mu U \tag{2.17}$$

[In the absence of sources, finite-energy solutions are necessarily trivial [6]; thus the potentials either vanish or are pure gauges.] A closed expression for this solution has not been given; only a formula perturbative in the source is available. So that we can speak of orders of perturbation, we shall take the source to be $O(Q)$ where Q is a convenient scale of magnitude for the source. [For example, Q can be an overall factor.] This solution is most economically presented by first transforming out of the Abelian frame with the gauge function U, occurring in (2.17).

$$\rho' = U\rho U^{-1} = U\frac{\sigma^3}{2i}U^{-1}q \tag{2.18a}$$

In the new frame, the vector potentials vanish with the source. Perturbative formulas for them are [1]

$$A'^{0} = \Phi + O(Q^3) \tag{2.18b}$$

$$\vec{A}' = \frac{1}{\nabla^2}[\Phi,\vec{\nabla}\Phi] + O(Q^4) \tag{2.18c}$$

$$\Phi = -\frac{1}{\nabla^2}\rho' \tag{2.18d}$$

Primes remind that the quantities are displayed in a non-Abelian frame. The gauge function U is not arbitrary but must be chosen so that the consistency condition (2.2) is satisfied. It is a consequence of that equation and of (2.18b) that we must have

$$[\Phi,\nabla^2\Phi] = 0 \tag{2.19}$$

The following is the temporal gauge equivalent to (2.18).

$$A'^{0} = 0 \tag{2.20a}$$

$$\vec{A}' = \vec{\nabla}\Phi t + (\tfrac{1}{2}t^2 + \frac{1}{\nabla^2})[\Phi,\vec{\nabla}\Phi] + O(Q^3) \tag{2.20b}$$

The electric field is $O(Q)$; the magnetic field, $O(Q^2)$.

$$\vec{E}' = -\vec{\nabla}\Phi - t[\Phi,\vec{\nabla}\Phi] + O(Q^3) \tag{2.20c}$$

$$\vec{B}' = \vec{\nabla}^{-1} \times [\Phi,\vec{\nabla}\Phi] + O(Q^3) \tag{2.20d}$$

$$\vec{\nabla}^{-1} = \vec{\nabla}/\nabla^2$$

[The time dependence in (2.20) is of course a consequence of the gauge choice, as comparison with (2.18) shows.] In the primed frame, the solution appears similar to the Coulomb one (2.9) or (2.10), save that the non-vanishing commutator $[\Phi,\vec{\nabla}\Phi]$ prevents the expressions from closing. Hence we call the above a "non-Abelian Coulomb" solution to contrast it with the "Abelian Coulomb" discussed at the outset. The energy further exhibits similarities with the Abelian Coulomb case. The formula to lowest order in Q follows from (2.3), (2.18d) and (2.20c).

$$\mathcal{E} = \tfrac{1}{2}\int \rho'_a \frac{-1}{\nabla^2}\rho'_a + O(Q^4)$$

$$= \frac{1}{8\pi}\int d\vec{r}d\vec{r}'\,\frac{\rho'_a(\vec{r})\rho'_a(\vec{r}')}{|\vec{r}-\vec{r}'|} + O(Q^4) \tag{2.21}$$

A specific example of a non-Abelian Coulomb solution is given when the source in the Abelian frame is spherically symmetric.

$$\rho_a = \delta_{a3} q(r) \tag{2.22a}$$

One then verifies that (2.19) is satisfied with the charge density in the radial frame;

$$\rho'_a = \hat{r}^a q(r) \tag{2.22b}$$

i.e. U is the gauge transformation which rotates the third axis into the radial axis. A further interesting feature is that the present solution carries less energy than the corresponding Coulomb one [7].

$$\mathcal{E} = \frac{1}{8\pi} \int d\vec{r}\, d\vec{r}\,' \frac{q(r) q(r')}{|\vec{r}-\vec{r}\,'|} \hat{r}\cdot\hat{r}' + O(Q^4) < \mathcal{E}_c \tag{2.22c}$$

The fourth and last solution, that I mention in this sub-Section, generalizes the static non-Abelian Coulomb in a time-dependent fashion, quite similarly to the way that the second solution, Eqs. (2.13), generalizes the Abelian Coulomb. It is constructed by choosing an arbitrary gauge transformation U and transforming the source once again.

$$\rho'' = U \rho' U^{-1} \tag{2.23}$$

We use double primes to distinguish this source from ρ -- the source in the Abelian frame -- and from ρ' -- the source in the gauge transformed frame where the non-Abelian Coulomb solution has a simple perturbative expansion, see (2.18) or (2.20). Next we take the regular solution of Poisson's equation

$$\Phi = -\frac{1}{\nabla^2} \rho'' \tag{2.24}$$

and build the time-dependent solution perturbatively in Φ, in the temporal gauge [5].

$$A''^{o} = 0 \tag{2.25a}$$

$$\vec{A}'' = \vec{\nabla}\Phi t + \left(\frac{t^2}{2} + \frac{1}{\nabla^2}\right)\left([\Phi, \vec{\nabla}\Phi] - \vec{\nabla}^{-1}[\Phi, \nabla^2 \Phi]\right) + O(Q^3) \tag{2.25b}$$

One readily computes the electric field, which is $O(Q)$.

$$\vec{E}'' = -\vec{\nabla}\Phi - t\left([\Phi, \vec{\nabla}\Phi] - \vec{\nabla}^{-1}[\Phi, \nabla^2 \Phi]\right) + O(Q^3) \tag{2.25c}$$

The $O(Q^2)$ magnetic field

$$\vec{B}'' = \vec{\nabla}^{-1} \times [\Phi, \vec{\nabla}\Phi] + O(Q^3) \tag{2.25d}$$

is of the same form as in the static non-Abelian Coulomb solution, (2.20d), which is here included when U=I. Just as in the Abelian situation, by continuously changing I to U, we obtain a continuous deformation of the static non-Abelian Coulomb into its time-dependent generalization. In both cases the magnetic field retains the same form during the deformation. Once again, the $O(Q^2)$ energy is given by a Coulombic formula, as follows from (2.3), (2.24) and (2.25c).

$$\mathcal{E} = \tfrac{1}{2}\int \rho''_a \frac{-1}{\nabla^2} \rho''_a + O(Q^4)$$
$$= \frac{1}{8\pi}\int d\vec{r}d\vec{r}\,' \frac{\rho''_a(\vec{r})\rho''_a(\vec{r}\,')}{|\vec{r}-\vec{r}\,'|} + O(Q^4) \qquad (2.26)$$

In Section 3, we show that also the above energy can lie below the corresponding energy of the static solution (2.22c) by an arbitrary amount.

The similarities between the four solutions should be apparent. Indeed, if for different gauge frames a common [unprimed] notation is used, a master formula which presents all four may be given [5]. Define first the vector $\vec{C}$.

$$\vec{C} = [\Phi, \vec{\nabla}\Phi] \qquad \Phi = \frac{-1}{\nabla^2}\rho \qquad (2.27)$$

Then in the temporal gauge, set

$$A^o = 0$$
$$\vec{A} = \frac{-1}{\nabla^2}\,[\vec{\nabla}\rho t + (\tfrac{1}{2}t^2 + \frac{1}{\nabla^2})\vec{\nabla}\times\vec{\nabla}\times\vec{C}] \qquad (2.28)$$

The static, Abelian Coulomb has vanishing $\vec{C}$, while vanishing $\vec{\nabla}\times\vec{C}$ leads to its time-dependent generalization as is seen from (2.12) and (2.13). Furthermore, according to (2.19) and (2.20), vanishing $\vec{\nabla}\cdot\vec{C}$ corresponds to the static, non-Abelian Coulomb solution while its time-dependent generalization (2.25) has no restrictions on $\vec{C}$. For the first two solutions (2.28) is exact; for the last two it is accurate up to $O(Q^2)$. The $O(Q^2)$ formula

$$\mathcal{E} = \tfrac{1}{2}\int \rho_a \frac{-1}{\nabla^2}\rho_a + O(Q^4) \qquad (2.29)$$

gives the complete energy for the exact solutions, and the $O(Q^2)$ contribution to the perturbative ones. For the two static solutions, the quantity in (2.29) is stationary against variations of ρ_a which preserve its length [gauge transformations] [8]. The Coulomb solution is seen to maximize (2.29).

2.3 Sources with Critical Strength

When the source strength Q increases, the previous solutions continue to be present. For the Abelian Coulomb, and its time-dependent generalization, the closed expressions given above hold for arbitrary Q. For the non-Abelian Coulomb with its time dependent generalization, one must calculate perturbatively terms higher order in Q; a tedious procedure with unknown convergence properties. Alternatively one can do numerical computations.

Furthermore as Q increases, solutions appear which require a critical, minimal source strength to support them. Very little is known about these, and the numerical method is presently the only effective means of investigation. We review one such example [1,9].

When the source is radially symmetric, as in (2.22a), we have the spherically symmetric Abelian Coulomb solution. Also by passing to the radial frame (2.22b) we can exhibit the perturbative non-Abelian Coulomb solution. By iterating Eqs. (2.18) a few orders in Q, it is found that the form of the potentials remains within the following Ansatz.

$$A^{o} = \frac{\hat{r}\cdot\vec{\sigma}}{2i}\ \frac{1}{r}\ f(r/r_o) \tag{2.30a}$$

$$\vec{A} = \frac{\hat{r}\times\vec{\sigma}}{2i}\ \frac{1}{r}\ [a(r/r_o)-1] \tag{2.30b}$$

Here r_o is a length scale. In this sub-Section we shall always remain in the radial frame,

$$\rho_a = \frac{\hat{r}^a}{r_o^3}\ q(r/r_o) \tag{2.31}$$

hence primes on the potentials are dropped. The above Ansatz is postulated for the complete static solution and the mode functions satisfy the following non-linear differential equations, which are all that remain of Gauss' and Ampère's laws.

$$-f'' + \frac{2a^2}{x^2}\ f = xq \tag{2.32a}$$

$$-a'' + \frac{a^2-1-f^2}{x^2}\ a = 0 \tag{2.32b}$$

All functions depend only on $x=r/r_o$, and the dash indicates differentiation with respect to that variable. More general radially symmetric Ansätze can be given, but it has been proven that static, radial solutions necessarily fall into the above restrictions [1]. [We emphasize that the Abelian Coulomb solution does not lie within the Ansatz (2.30), and cannot be found in the solutions to (2.32); in the radial frame, the Abelian Coulomb solution is not radially symmetric.] Requiring finiteness of the energy

$$\mathcal{E} = \frac{4\pi}{r_o} \int_0^{\infty} dx[(a')^2+\frac{1}{2x^2}(a^2-1)^2+\tfrac{1}{2}(f')^2+\frac{1}{x^2}f^2a^2] \tag{2.33}$$

-- the above is the form that (2.3) takes within the Ansatz (2.30) -- imposes boundary conditions at the origin and at infinity. At the origin the potentials must vanish rapidly: $f(0)=0$, $a(0)=1$, $A^o(0)=0$, $\vec{A}(0)=0$. At infinity two types of behavior are allowed: type I, where the potentials vanish as in the origin; type II, where the vector potential tends to a nontrivial pure gauge, $a(\infty)=-1$, $\vec{A} \underset{r\to\infty}{\to} i\frac{\hat{r}\times\vec{\sigma}}{r} = -(i\vec{\sigma}\cdot\hat{r})\vec{\nabla}(-i\sigma\cdot\hat{r})$. The type I solution is the previously perturbatively encountered non-Abelian Coulomb $[a=1+0(Q^2)$, hence a never equals $-1]$. The type II is a new, non-perturbative solution.

Numerical computation confirms the above, with the further surprise that type II comes in two branches, once Q exceeds a critical magnitude [1]. Hence we call this the "bifurcating" solution. Fig. 1 shows a plot of the energy versus source strength for solutions with a delta-shell charge density.

$$\rho_a = \frac{\hat{r}^a}{r_o^2} Q\delta(r-r_o) \qquad (2.34)$$

The Coulomb parabola [which does not lie within the Ansatz] is also exhibited for comparison. Note that the non-Abelian Coulomb [type I] carries lower energy than the Abelian Coulomb for all Q, even outside the perturbative regime. The bifurcation point where the two type II solutions first occur is found numerically to be Q=5.835. In Figs. 2,3 and 4 the mode functions f and a are displayed for the various solutions. All figures are taken from Ref. [1].

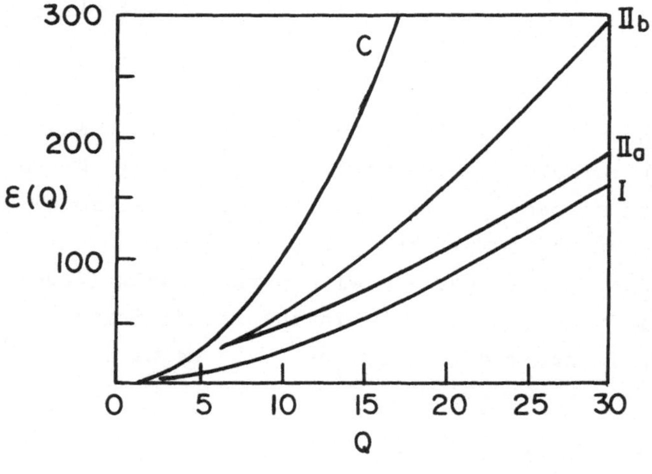

Fig.1 Energy, in units of $2\pi/r_o$ as a function of Q for a delta-shell source of strength Q. The curve C is the Abelian Coulomb parabola. The curve I is the non-Abelian Coulomb solution. Curves IIa and IIb are the two branches of the bifurcating solution. The bifurcation point occurs at Q=5.826.

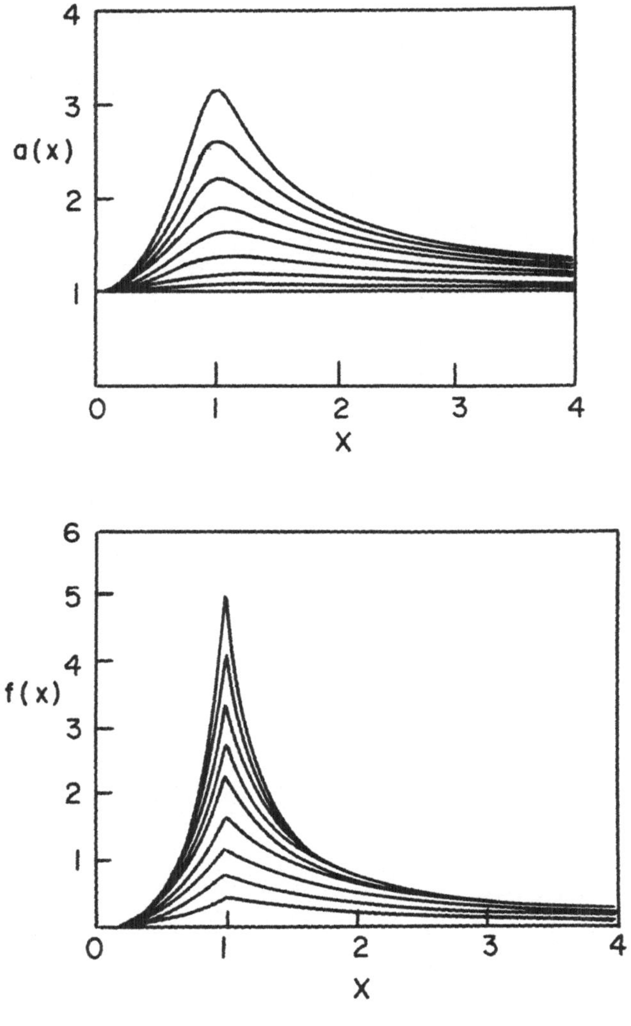

Fig.2 Profiles of the functions a and f for the type I solution. Starting from the lowest curves the values of Q in a delta-shell source are 1.41, 2.53, 4.04, 6.43, 10.05, 14.18, 19.93, 23.38 and 41.72.

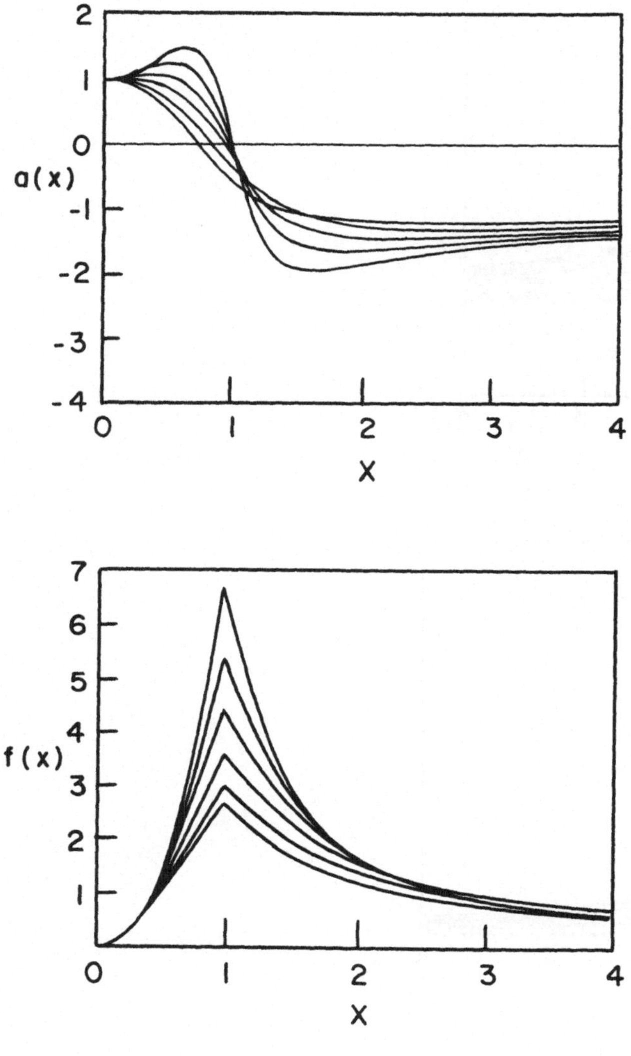

Fig.3 Profiles of the functions a and f for the (a)(lower) branch of the type II bifurcating solution with a delta-shell source. Q=5.86, 6.44, 8.09, 11.05, 15.71 and 23.19. Correspondence between the individual curves and these values of Q is established by the fact that, as Q increases, so do a"(0) and f(1).

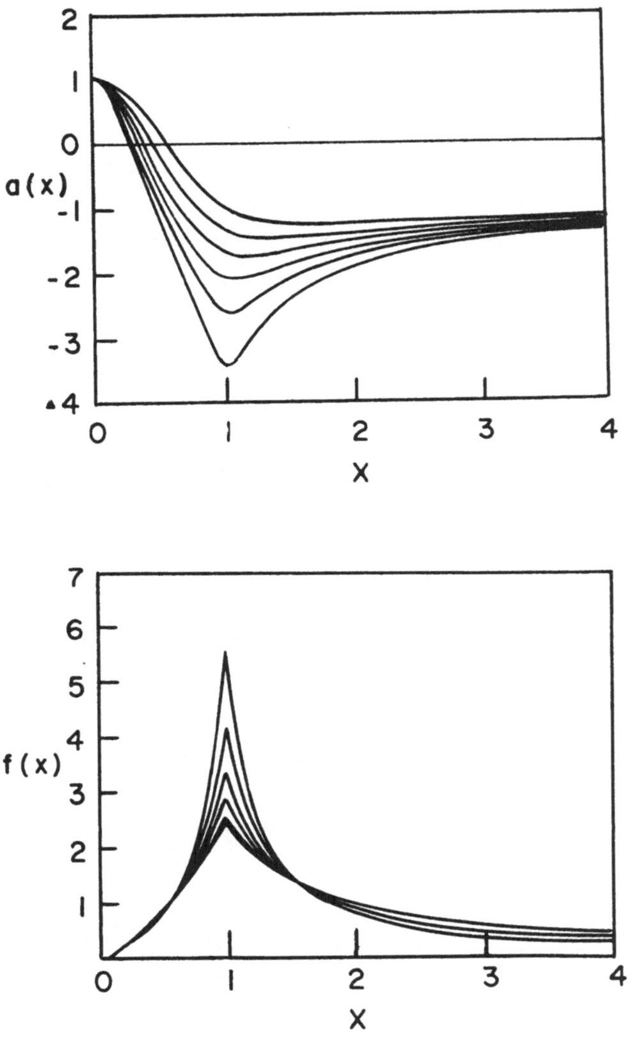

Fig.4 Profiles of the functions a and f for the (b)(upper) branch of the type II bifurcating solution with a delta-shell source. Q=6.53, 8.61, 12.10, 17.85, 28.06 and 49.16. For these curves, as Q increases a"(0) decreases and f(1) increases.

3. Stability in Dynamical Systems

3.1 Review of the General Theory

We consider a time-translation invariant system whose equations of motion for the 2N dynamical variables P_n and Q_n, $n=1,...,N$, can be obtained from a Hamiltonian $H(P,Q)$, which is also the conserved energy $\mathcal{E}$.

$$\dot{P}_n = -\frac{\partial H(P,Q)}{\partial Q_n}$$

$$\dot{Q}_n = \frac{\partial H(P,Q)}{\partial P_n} \qquad (3.1)$$

A static solution , one for which $\dot{P}$ and $\dot{Q}$ vanish, is a critical point of H, and vice versa, stationary points of the energy define static solutions. [An over-dot means differentiation with respect to time.]

We wish to ascertain whether a static solution $\{P^{(s)}, Q^{(s)}\}$ is stable. "Stable" by definition will mean the following: Take a configuration of the form $\{P^{(s)}+\delta P, Q^{(s)}+\delta Q\}$, substitute in (3.1) and linearize about $\{P^{(s)}, Q^{(s)}\}$ to obtain linear equations for the fluctuating quantities $\{\delta P,\delta Q\}$. When the linear equations produce exponential growth in time for the fluctuations, the solution is unstable; otherwise, it is stable. In other words, for stable motion the small quantities $\{\delta P,\delta Q\}$ fluctuate harmonically in time with real frequency, while complex frequencies signal instability.

The above criterion for stability is also in accord with quantum-mechanical ideas. The first quantum correction to the energy of a state involves the fluctuation frequencies. That quantity must be real for the state to be quantum-mechanically stable.

Note that growth in time of the fluctuations smaller than exponential, say polynomial, is not a sign of instability. In such a circumstance, the eigenfrequencies are degenerate, but still real, and the quantal energy remains real.

An intuitively appealing idea is that stability should be connected with minimizing the energy: the static solution which is a stationary point should also be a [local] minimal point. More precisely, the minimality condition is the requirement that only non-negative eigenvalues characterize the quadratic Hamiltonian matrix, $\mathcal{H}$, defined by expanding $H(P,Q)$ about $\{P^{(s)},Q^{(s)}\}$ and retaining quadratic terms in

$\{\delta P,\delta Q\}$. [Linear terms vanish since the expansion is about a critical point.]

$$H(P,Q) = H(P^{(s)},Q^{(s)}) + \tfrac{1}{2}\delta P_n T_{nm}\delta P_m + \delta P_n G_{nm}\delta Q_m + \tfrac{1}{2}\delta Q_n V_{nm}\delta Q_m + \ldots$$

$$= H(P^{(s)},Q^{(s)}) + \tfrac{1}{2}\tilde{X}\mathcal{H}X + \ldots \qquad (3.2)$$

$$\mathcal{H} = \begin{pmatrix} T & G \\ \tilde{G} & V \end{pmatrix} \qquad X = \begin{pmatrix} \delta P \\ \delta Q \end{pmatrix} \qquad (3.3)$$

[The tilde indicates transposition.] The minimality condition demands

$$\det(\mathcal{H} - \lambda I) = 0 \rightarrow \lambda \geq 0 \qquad (3.4)$$

In fact minimality is a sufficient condition for stability -- a result, known as Dirichlet's theorem, which will become apparent below -- but by no means is it a necessity [10]. There are indeed familiar physical systems [tops, gyroscopes, planetary configurations] which are stable, even though their energy is not locally minimal. To derive a more general criterion we expand (3.1) around the static solution and find

$$\mathcal{H} X = i\eta\dot{X} \qquad (3.5)$$

$$\eta = \begin{pmatrix} 0 & -iI \\ iI & 0 \end{pmatrix} \qquad (3.6)$$

By making a monochromatic <u>Ansatz</u> for X,

$$X = e^{-i\omega t}x \qquad (3.7)$$

we recognize the [constant] x as symplectic eigenvectors of $\mathcal{H}$ with simplectic eigenvalue ω.

$$\mathcal{H} x = \omega\eta x \qquad (3.8)$$

It is clear that our definition of stability requires the ω's be real; this is known as Liapunov's theorem [10].

$$\det(\mathcal{H} - \omega\eta) = 0 \rightarrow \omega \text{ real} \qquad (3.9)$$

The point is that (3.9) is in general different from (3.4) and can be satisfied when (3.4) fails.

If (3.8) is premultiplied by $x^\dagger$, where the dagger indicates transposition and complex conjugation,

$$x^\dagger \mathcal{H} x = \omega x^\dagger \eta x \qquad (3.10)$$

we see that the left-hand side is real, $\mathcal{H}$ being real symmetric, hence

Hermitian. Also $x^\dagger \eta x$ is real since η is Hermitian, and we conclude that ω can fail to be real only when $x^\dagger \mathcal{H} x$ and $x^\dagger \eta x$ vanish. So when $\mathcal{H}$ is positive definite, ω is real and Dirichlet's theorem is established: minimality implies stability. But of course ω may be real without $\mathcal{H}$ being positive definite.

One may consider η as a metric in the vector space of the x's. Then (3.8) is the condition that $x^\dagger \mathcal{H} x$ be stationary against variations of x which preserve the symplectic length $x^\dagger \eta x$. Instability can occur only when there are zero-length simplectic eigenvectors of $\mathcal{H}$. The eigenvalue equation in (3.9) is relevant to the program of diagonalizing $\mathcal{H}$ by simplectic matrices, just as the corresponding equation in (3.4) arises when diagonalizing with orthogonal matrices. [A matrix M is simplectic when $\tilde{M}\eta M = \eta$.]

The conditions (3.4) and (3.9) are clearly different, and no simple relationship exists between the two in the general case. In practice, we can specialize somewhat. Firstly, the kinetic energy matrix T in (3.2) and (3.3) is taken to be positive definite: with an appropriate definition of coordinates, we may choose it to be the identity. Secondly, the off-diagonal matrix G arising from mixed p-q terms in the Hamiltonian, which are frequently called gyroscopic or Coriolis terms, is always anti-symmetric, when the theory is derivable from a Lagrangian. The reason is that any symmetric piece in such velocity dependent forces corresponds to a total time-derivative in the Lagrangian and may be dropped. Thus we are led to a simpler form for $\mathcal{H}$.

$$\mathcal{H} = \begin{pmatrix} I & G \\ -G & V \end{pmatrix}$$
$$\tilde{G} = -G, \quad \tilde{V} = V \tag{3.11}$$

With this $\mathcal{H}$, the simplectic eigenvalue problem (3.9) reduces to

$$[(i\omega+G)(i\omega+G)+V]\delta Q = 0 \tag{3.12a}$$

and the stability condition becomes

$$\det\,(2i\omega G+G^2+V-\omega^2 I) = 0 \;\rightarrow\; \omega \;\; \text{real} \tag{3.12b}$$

To compare the above with the minimality condition (3.4), we impose that requirement not on $\mathcal{H}$, but on the equivalent matrix $\tilde{M}\mathcal{H} M$ which has the same positivity properties as $\mathcal{H}$. Here

$$M = \begin{pmatrix} I & -G \\ 0 & I \end{pmatrix}, \quad \tilde{M}\mathcal{H} M = \begin{pmatrix} I & 0 \\ 0 & G^2+V \end{pmatrix}$$

Then (3.4) becomes equivalent to

$$\det (G^2+V-\lambda I) = 0 \rightarrow \lambda \geq 0 \tag{3.13}$$

This is analogous to (3.12b), but an obvious difference exists when G is present. When the gyroscopic forces are absent, the two conditions coïncide and ω^2 may be identified with λ. In that case instability occurs only for imaginary ω. In the presence of gyroscopic terms, there exist stable static solutions which do not minimize the energy, while instability can exist with complex ω. When $\mathcal{H}$ is as in (3.11), the condition for instability, $x^\dagger \eta x=0$, is equivalent to

$$\mathrm{Re}\omega = i \frac{\delta Q_n^* G_{nm} \delta Q_m}{\delta Q_n^* \delta Q_n} \tag{3.14}$$

We shall use the phrase "gyroscopic stability" when we wish to distinguish this form of stability from the more familiar "energetic stability". A hint for gyroscopic stability occurs when we can find arbitrarily close to a static solution oscillatory fluctuations that lower the energy. As we shall show, such configurations exist in the Yang-Mills theory. Instability would be indicated when there are, arbitrarily close to the static solution, time-dependent solutions which decrease the energy and grow exponentially in time.

To conclude this review of stability theory, let us remark that although we discussed Dirichlet's sufficient condition in terms of the energy constant of motion, a similar criterion can be formulated by reference to other constants of motion. This generalization is useful when analyzing solutions invariant with respect to the symmetry transformation which is associated with the constant in question [10].

3.2 Stability Analysis for Yang-Mills Theory

We turn now to the stability analysis of static solutions for the Yang-Mills equations; but before we use the ideas sketched above, we must recognize that there are two ways in which the Yang-Mills field theory differs from the simple Hamiltonian. Firstly, rather than 2N degrees of freedom, there is an infinite number. This causes matrices to be replaced by differential operators, summations by integrations, etc., thus raising questions of convergence and uniformity. We shall not concern ourselves with this complication, even though there will be occasion to refer to it in the course of our development. Secondly, the Hamiltonian formulation now has constraints. This has already been dealt with in Section 2. Here we observe that the small fluctuation equations, which follow from (2.6) by linearizing around a static solution A_a^μ, are

$$0 = \vec{\mathcal{D}}_{ab}\cdot\delta\vec{E}_b + \varepsilon_{abc}\vec{E}_b\cdot\delta\vec{A}_c \tag{3.15a}$$

$$\mathcal{D}^{o}_{ab}\delta\vec{E}_b = \vec{\mathcal{D}}_{ab}\times\delta\vec{B}_b+\varepsilon_{abc}\vec{E}_b\delta A^o_c-\varepsilon_{abc}\vec{B}_b\times\delta\vec{A}_c \tag{3.15b}$$

$$\delta\vec{E}_a = -\mathcal{D}^o_{ab}\delta\vec{A}_b-\vec{\mathcal{D}}_{ab}\delta A^o_b \tag{3.15c}$$

$\delta\vec{B}_a$ is short-hand for $\vec{\mathcal{D}}_{ab}\times\delta\vec{A}_b$. The quadratic energy, $\mathcal{E}^{(2)}$, obtained by expanding (2.3) around the static solution and by using Gauss' law constraint for the fluctuations, is precisely of the form (3.11).

$$\mathcal{E}^{(2)} = \tfrac{1}{2}\int d\vec{r}\{(\delta\vec{E}_a)^2+2\delta E^i_a(\varepsilon_{acb}A^o_c)\delta A^i_b + (\delta\vec{B}_a)^2-\delta A^i_a(\varepsilon^{ikj}\varepsilon_{acb}B^k_c)\delta A^j_b\} \tag{3.16}$$

In particular, an anti-symmetric gyroscopic term is present.

$$G = \delta^{ij}\delta(\vec{r}-\vec{r}\,')\varepsilon_{abc}A^o_c \tag{3.17}$$

Eqs. (3.15) can also be obtained by taking (3.16) to be the quadratic Hamiltonian, and varying it subject to the constraint (3.15a) which is implemented with the help of a Lagrange multiplier δA^o_a. In other words, unconstrained variations are performed on

$$\bar{\mathcal{E}}^{(2)} = \mathcal{E}^{(2)} - \int d\vec{r}\delta A^o_a(\vec{\mathcal{D}}_{ab}\cdot\delta\vec{E}_b+\varepsilon_{abc}\vec{E}_b\cdot\delta\vec{A}_c) \tag{3.18}$$

and Eqs. (3.15a), (3.15b), and (3.15c) emerge respectively as

$$0 = -\frac{\delta\bar{\mathcal{E}}^{(2)}}{\delta(\delta A^o_a)} \tag{3.19a}$$

$$\partial_t\delta\vec{E}_a = \frac{\delta\bar{\mathcal{E}}^{(2)}}{\delta(\delta\vec{A}_a)} \tag{3.19b}$$

$$\partial_t\delta\vec{A}_a = -\frac{\delta\bar{\mathcal{E}}^{(2)}}{\delta(\delta\vec{E}_a)} \tag{3.19c}$$

With a monochromatic Ansatz for the time-dependence, the above take on the form of a symplectic eigenvalue problem, equivalent to stationarizing $\mathcal{E}^{(2)}$, subject to the constraint that the symplectic length of $\begin{pmatrix}-\delta\vec{E}_a\\ \delta\vec{A}_a\end{pmatrix}$ be fixed, and subject to the constraint of Gauss' law for the fluctuations. In short, the Yang-Mills model is seen to fit the general theory quite nicely.

Let us comment on some of the properties of the small fluctuation equations (3.15). An integrability condition follows from (3.15b). By taking the covariant divergence, one finds that the infinitesimal version of (2.2) must be satisfied.

$$\varepsilon_{abc}\delta A_b^0 \rho_c = 0 \tag{3.20}$$

Also *vice versa*: (3.20) and the integrability condition on (3.15b) imply, together with (3.15c), that the covariant time-derivative of the right-hand side in (3.15a) vanishes.

Eqs. (3.15) possess a local gauge invariance.

$$\delta\vec{E}_a \to \delta\vec{E}_a - \varepsilon_{abc}\vec{E}_b\theta_c \tag{3.21a}$$

$$\delta A_a^0 \to \delta A_a^0 - \partial_t\theta_a - \varepsilon_{abc}A_b^0\theta_c \tag{3.21b}$$

$$\delta\vec{A}_a \to \delta\vec{A}_a + \vec{\nabla}\theta_a - \varepsilon_{abc}\vec{A}_b\theta_c \tag{3.21c}$$

Here θ_a is a local function which must be parallel to the source.

$$\varepsilon_{abc}\theta_b\rho_c = 0 \tag{3.22}$$

[There is also a gauge covariance: a gauge transformation on the background fields is compensated by a homogeneous gauge transformation on the fluctuating quantities. We shall not make use of this property.]

It is clear from (3.20) and (3.22) that the external charge density defines a direction in group space which we can call the "electromagnetic" direction, while the orthogonal directions can be termed "charged". Thus A_a^0, δA_a^0, θ_a and ρ_a all lie in the electromagnetic direction and vanish in the charged direction. This reduces the allowed gauge transformations, in that the last term in (3.21b) must vanish. Observe also that the gyroscopic term (3.17) affects only the charged direction; the electromagnetic fluctuations are free of gyroscopic terms.

It is possible to derive a gauge invariant fluctuation equation in the following way. The quantity

$$\begin{aligned}\vec{e}_a &= \delta\vec{E}_a + \varepsilon_{abc}A_b^0\delta\vec{A}_c \\ &= -\partial_t\delta\vec{A}_a - \vec{\mathcal{D}}_{ab}\delta A_b^0\end{aligned} \tag{3.23}$$

is gauge invariant with respect to the gauge transformations (3.21). By taking a covariant time-derivative of (3.15b) we arrive after some steps at

$$\mathcal{D}^0_{ab}\mathcal{D}^0_{bc}\vec{e}_c + \vec{\mathcal{D}}_{ab}\times\vec{\mathcal{D}}_{bc}\times\vec{e}_c - \varepsilon_{abc}\vec{B}_b\times\vec{e}_c = 0 \tag{3.24a}$$

[This is most readily obtained in the $\delta A^0 = 0$ gauge, which can always be achieved with the transformation (3.21b).] With a monochromatic

<u>Ansatz</u>

$$\vec{e}_a = e^{-i\omega t}\vec{\chi}_a \tag{3.24b}$$

(3.24a) may be written as

$$[i\omega\delta_{ab}+\varepsilon_{abm}A_m^o]\ [i\omega\delta_{bc}+\varepsilon_{bcn}A_n^o]\chi_c^i + V_{ac}^{ij}\chi_c^j = 0 \tag{3.24c}$$

$$V_{ac}^{ij} = \varepsilon^{ikm}\mathcal{D}_{ab}^m\mathcal{D}_{bc}^n\varepsilon^{nkj} - \varepsilon^{ikj}\varepsilon_{abc}B_b^k \tag{3.24d}$$

We see that (3.24c) is precisely of the form (3.12a), again showing that the Yang-Mills theory follows the analysis described in sub-Section 3.1.

Eqs. (3.24) are gauge invariant and involve the unconstrained $\vec{e}_a$. It is remarkable that such equations can be derived; the possibility to do so is intimately linked with the existence of an external charge density which defines a direction with respect to which the small fluctuations are constrained by (3.20) [5].

<u>3.2.1 Abelian Coulomb Solution</u>

For the Abelian Coulomb solution, the general stability theory is easily applied. The small oscillation equations are best presented by introducing complex quantities in the charged directions 1 and 2.

$$\delta\vec{E} = \frac{1}{\sqrt{2}}(\vec{E}_1+i\vec{E}_2)$$

$$\delta\vec{A} = \frac{1}{\sqrt{2}}(\vec{A}_1+i\vec{A}_2)$$

$$\vec{e} = \frac{1}{\sqrt{2}}(\vec{e}_1+i\vec{e}_2) \tag{3.25}$$

Eq. (3.24a) in the electromagnetic direction, 3, decouples completely.

$$\partial_t^2\vec{e}_3 + \vec{\nabla}\times\vec{\nabla}\times\vec{e}_3 = 0 \tag{3.26}$$

while in the charged direction we have simply

$$(\partial_t+i\phi)^2\vec{e} + \vec{\nabla}\times\vec{\nabla}\times\vec{e} = 0$$
$$\nabla^2\phi = -q \tag{3.27}$$

The electromagnetic fluctuations are free; the charged ones describe the motion of charged vector mesons in an external electric field with a potential ϕ [11].

Detailed analysis of the equations can be performed in frequency space. Note that the electromagnetic equation involves ω^2 as an eigenvalue of a Hermitian operator, hence it is real. Only the issue remains whether ω^2 is positive or negative. In the charged equation there appears $(\omega-\phi)^2$ and ω can be complex; it is not related to the eigenvalue of a Hermitian operator. This difference reflects the fact previously remarked upon: in the electromagnetic direction there are no gyroscopic terms, hence stability is equivalent to minimality. In the charged direction, gyroscopic terms are present; they are responsible for the more complicated equation.

The electromagnetic fluctuations are obviously stable. Those in the charged directions are stable in the absence of the external potential and by continuity they remain stable for a sufficiently small external potential [3]. As the external charge density increases in strength, an instability is expected to appear. This is not the instability of the Klein-Gordon equation in a 1/r [Coulomb] potential, which has previously been remarked upon, and which is a consequence of the [presumably unphysical] singularity at the origin [12]. In our expamles the potentials are non-singular. Instead it is the instability of the Klein-Gordon equation in a strong external field [13].

In spite of stability for weak sources, we expect as a consequence of the gyroscopic terms to find modes which, though oscillatory, lower the energy. These can be readily exhibited, without passing to frequency space. We remain with the first-order equations (3.15), and seek a solution with $\delta\vec{B}_a=0$. In that case the charged portions of (3.15), with an Abelian Coulomb solution as the background field, reduce to

$$0 = \vec{\nabla}\cdot\delta\vec{E} - i\vec{\nabla}\phi\cdot\delta\vec{A} \tag{3.28a}$$

$$0 = (\partial_t + i\phi)\delta\vec{E} \tag{3.28b}$$

$$\delta\vec{E} = -(\partial_t + i\phi)\delta\vec{A} \tag{3.28c}$$

$$\delta\vec{B} = \vec{\nabla}\times\delta\vec{A} = 0 \tag{3.28d}$$

The solution of (3.28b) and (3.28c) is

$$\delta\vec{A} = [\vec{a}_o(\vec{r})+t\vec{a}_1(\vec{r})]\exp-it\phi(\vec{r}) \tag{3.29a}$$

$$\delta\vec{E} = -\vec{a}_1(\vec{r})\exp-it\phi(\vec{r}) \tag{3.29b}$$

and to satisfy (3.28a) and (3.28d) we must have

$$\vec{a}_o = \vec{\nabla}\theta \tag{3.29c}$$

$$\vec{a}_1 = -i\theta\vec{\nabla}\phi-i\vec{\nabla}^{-1}(\theta q) \tag{3.29d}$$

where θ is an arbitrary function. Finally there is one more condition: $\nabla^{-1}(\theta q)$ must be parallel to $\vec{\nabla}\phi$, which can be easily achieved, for example by setting $\theta q=\nabla^2 F(\phi)$, where F is arbitrary. Thus equations (3.28) can be satisfied in terms of one function.

The quadratic energy (3.16) is seen to be negative [14].

$$\mathcal{E}^{(2)} = \int [(q\theta)\ \frac{-1}{\nabla^2}\ (q\theta^*) - |\theta|^2 (q)\ \frac{-1}{\nabla^2}\ (q)]$$
$$= \frac{-1}{8\pi} \int d\vec{r}d\vec{r}'\ \frac{q(\vec{r})q(\vec{r}')}{|\vec{r}-\vec{r}'|}\ |\theta(\vec{r})-\theta(\vec{r}')|^2 \qquad (3.30)$$

This rather peculiar fluctuation gives evidence that the Coulomb solution is gyroscopically stable since the energy decreases below its Coulomb value [5]. Note from (3.29), the linear growth of the fluctuation with time, which however does not produce instability. Because the frequency is position dependent, it is not clear how to locate this mode among superpositions of functions with definite frequency. Nevertheless we have encountered it already! It is merely the time-dependent generalization of the Abelian Coulomb solution, Eqs. (2.12)-(2.16), when the latter is brought arbitrarily close to the static solution [15]. The energy formula (3.30) can now be recognized as the $O(\theta^2)$ contribution to (2.16), when the source ρ' is a gauge transformation, with gauge function θ, of ρ the source in the standard frame.

3.2.2 Non-Abelian Coulomb Solution

The non-Abelian Coulomb solution, Eqs. (2.18), follows in many respects, at least for weak sources, the behavior of the Abelian Coulomb solution. [Only the weak source regime is amenable to analytic treatment, since our formulas are given by a source strength power series.] The stability equations are now highly coupled, and have not been solved. However, by continuity with the sourceless problem one expects stability for weak sources [3]. Moreover, one can show that this again must be an instance of gyroscopic stability, since the energy is lowered by the time-dependent generalization, presented in Eqs. (2.23)-(2.26), which can be taken arbitrarily close to the non-Abelian Coulomb solution. This is achieved by making ρ'', the source for the time-dependent solution, to be an infinitesimal gauge transformation of ρ', the source in the non-Abelian Coulomb solution. The energy can then be computed from (2.26). The details are the following. Set

$$\rho''_a = (1-\tfrac{1}{2}\theta_b\theta_b)\ \rho'_a - \varepsilon_{abc}\theta_b\rho'_c \qquad (3.31)$$

This assures that ρ''_a is a gauge transformation of ρ'_a taken through second order in θ_a, which lies only in the charged direction.

$$\theta_a(\vec{r})\rho'_a(\vec{r}) = 0 \tag{3.32}$$

It now follows from (2.26) that the $O(Q^2)$ energy is

$$\begin{aligned}\mathcal{E} = &\frac{1}{8\pi}\int d\vec{r}d\vec{r}'\,\frac{\rho'_a(\vec{r})\rho'_a(\vec{r}')}{|\vec{r}-\vec{r}'|} \\ &-\frac{1}{4\pi}\int d\vec{r}d\vec{r}'\,\frac{\rho'_a(\vec{r})\rho'_a(\vec{r}')}{|\vec{r}-\vec{r}'|}\,\varepsilon_{abc}\theta_c(\vec{r}') \\ &-\frac{1}{16\pi}\int d\vec{r}d\vec{r}'\,\frac{\rho'_a(\vec{r})\rho'_a(\vec{r}')}{|\vec{r}-\vec{r}'|}\,[\theta_b(\vec{r})-\theta_b(\vec{r}')]^2 \\ &-\frac{1}{8\pi}\int d\vec{r}d\vec{r}'\,\frac{\rho'_a(\vec{r})\rho'_a(\vec{r}')}{|\vec{r}-\vec{r}'|}\,\theta_a(\vec{r}')\theta_b(\vec{r}) + O(Q^4)\end{aligned} \tag{3.33}$$

The first term is the $O(Q^2)$ non-Abelian Coulomb energy. The second may also be written as $\int d\vec{r}\Phi_a(\vec{r})\varepsilon_{abc}\nabla^2\Phi_b(\vec{r})\theta_c(r)$, whence it is seen to vanish due to (2.19). The remaining two terms give the energy of the fluctuation. Unlike in (3.30), one cannot determine the sign by inspection, but after some straightforward manipulations, one can show that in the generic spherically symmetric case the terms are negative [5].

3.2.3 Bifurcating Solutions

The bifurcating solutions, described in Section 2.3, exist only for sufficiently strong sources. Consequently, we have no closed-form expressions to analyze; yet precisely because there is a bifurcation, we can say a considerable amount without explicit computation. Consider first a solution to the static Yang-Mills equations for a definite source ρ-- Eqs. (2.6) with the left-hand side of (2.6b) and (2.6c) set to zero. Next imagine changing the source strength slightly, $\rho\to\rho+\delta\rho$, and looking for a new static solution. If the new solution is regularly related to the old one, the increments in the Yang-Mills fields will satisfy linear equations which are of the same form as the fluctuation equations (3.15), except that $\delta\rho$ occurs in the left-hand side of (3.15a) and time-derivatives are absent in (3.15b) and (3.15c). However, if we are at the bifurcation point, it must be impossible to solve these equations, and this happens if the homogeneous system has a non-trivial solution. In this way we arrive at the important observation that at the bifurcation point the stability equations have a zero-eigenvalue mode, and vice versa: a zero-eigenvalue mode may signal bifurcation [or generalizations thereof] in the static solutions viewed as functionals of the source.

It is possible to use general properties of bifurcation phenomena to give a description of the Yang-Mills theory near the bifurcation point. The discussion is complicated by a proliferation of equations and indicies. Therefore we present here an analysis of a simplified, one-component model, in order to exemplify the general theory. The Yang-Mills problem follows completely the features of the example, and details are presented elsewhere [5].

Let us consider a non-linear field equation for the field $\Phi(t,\vec{r})$ in the presence of an external, static source $\rho(\vec{r})$.

$$\left(\frac{\partial^2}{\partial t^2} - \nabla^2\right)\Phi + U'(\Phi) = -\rho \tag{3.34}$$

Here U is a potential for the field, and the dash indicates a differentiation with respect to argument. We suppose that a bifurcation occurs at $\rho = \rho_c$. By hypothesis ρ_c supports a unique static solution $\phi_c(\vec{r})$; with stronger source there exists more than one solution. Correspondingly, there is a real, normalized zero-eigenvalue mode $\psi(\vec{r})$ in the small fluctuation equations.

$$-\nabla^2\phi_c + U'(\phi_c) = -\rho_c \tag{3.35a}$$

$$\{-\nabla^2 + U''(\phi_c)\psi = 0 \tag{3.35b}$$

Let us now replace ρ by $\rho_c+\varepsilon\delta\rho$, where ε is a small parameter, chosen to be positive, which systematizes the study of the theory around the bifurcation point. It is then appropriate to expand the static field ϕ according to

$$\phi = \phi_c + \varepsilon^{\frac{1}{2}}c\psi + \varepsilon\delta\phi + \ldots \tag{3.36}$$

with c being a numerical factor which together with $\delta\phi$ is to be determined. Expansion of (3.34) shows that terms independent of ε as well as those of order $\varepsilon^{\frac{1}{2}}$ vanish by virtue of (3.35). The order ε equation leaves

$$\{-\nabla^2+U''(\phi_c)\}\delta\phi + \frac{1}{2}\, c\; U'''\,(\phi_c)\psi^2 = -\delta\rho \tag{3.37}$$

Eq. (3.35b) implies a consistency condition on (3.37).

$$\frac{1}{2}c^2 \int d\vec{r}U'''(\phi_c)\psi^3 = -\int d\vec{r}\psi\delta\rho \tag{3.38}$$

For generic $\delta\rho$, the right-hand side of (3.38) is non-vanishing. We shall further assume that the integral in the left-hand side also is non-vanishing.

$$\int d\vec{r}\, U'''(\phi_c)\psi^3 \neq 0 \qquad (3.39)$$

[This assumption is a prerequisite for the subsequent development. Although we have not checked its validity in our Yang-Mills application, the fact that our analysis of the bifurcation is verified in the explicit numerical results, see below, provides an a posteriori justification.] Thus (3.38) determines both the magnitude of c and the direction of the bifurcation.

$$c^2 = 2\frac{-\int d\vec{r}\psi\delta\rho}{\int d\vec{r}\, U'''(\phi_c)\psi^3} \qquad (3.40a)$$

Obviously the functional sign of $\delta\rho$ must be such that the right-hand is positive. [We seek real solutions, hence c in (3.36) must be real.] Two solutions for c are obtained

$$c = \pm\left|\frac{2\int d\vec{r}\ \psi\delta\rho}{\int d\vec{r}\, U'''(\phi_c)\psi^3}\right|^{\frac{1}{2}} \qquad (3.40b)$$

and consequently (3.36) shows that ϕ bifurcates around ϕ_c. Once the consistency condition (3.40) is satisfied, (3.37) may be solved for $\delta\phi$.

The energy around the bifurcation point also is readily determined. The energy of a static solution to (3.34) is

$$\mathcal{E} = \int d\vec{r}[\tfrac{1}{2}(\vec{\nabla}\phi)^2+U(\phi)+\rho\phi] \qquad (3.41a)$$

When this is differentiated with respect to ε we find

$$\frac{\partial\mathcal{E}}{\partial\varepsilon} = \int d\vec{r}\ \frac{\delta\mathcal{E}}{\delta\phi(\vec{r})}\frac{\partial\phi(\vec{r})}{\partial\varepsilon} + \int d\vec{r}\phi(\vec{r})\frac{\partial\delta\rho}{\partial\varepsilon} \qquad (3.41b)$$

The first term on the right-hand side vanishes, since a static solution stationarizes the energy, while in the second term we use our assumed form for ϕ and ρ. Thus we find

$$\frac{\partial\mathcal{E}}{\partial\varepsilon} = \int d\vec{r}(\phi_c+\varepsilon^{\frac{1}{2}}c\psi+\ldots)\delta\rho$$

$$\mathcal{E} = \mathcal{E}_c + \varepsilon\int d\vec{r}\phi_c\delta\rho+\frac{2}{3}\, c\varepsilon^{\frac{3}{2}}\int d\vec{r}\psi\delta\rho+\ldots \qquad (3.41c)$$

Since c can take on two different signs according to (3.40), the above shows that the energy bifurcates, with an energy difference rising as $\varepsilon^{\frac{3}{2}}$.

Finally we examine stability of the bifurcating solutions. The oscillatory modes associated with (3.44) staisfy

$$\Phi = \phi + e^{-i\omega_n t}\Psi_n \qquad (3.42a)$$

$$\{-\nabla^2 + U''(\phi)\}\Psi_n = \omega_n^2 \Psi_n \qquad (3.42b)$$

We concentrate on the mode Ψ_o, which at the bifurcation point is the zero-eigenvalue mode, $\phi=\phi_c$, $\Psi_o=\psi$, and examine what happens immediately above the bifurcation, where we may set

$$\phi = \phi_c + \epsilon^{\frac{1}{2}} c\psi + \ldots \qquad (3.43a)$$

$$\Psi_o = \psi + \delta\Psi_o \qquad (3.43b)$$

ω_o^2 is taken to be a small quantity, since in lowest order it vanishes. Inserting (3.43) into (3.42b) gives, to first order in small quantities,

$$\omega_o^2 \psi = \{-\nabla^2 + U''(\phi_c)\}\delta\Psi_o + c\epsilon^{\frac{1}{2}} U'''(\phi_c)\psi^2 \qquad (3.44a)$$

Eq. (3.35) implies a consistency condition on the above; this evaluates ω_o^2.

$$\omega_o^2 = c\epsilon^{\frac{1}{2}} \int d\vec{r}\, U'''(\phi_c)\psi^3 \qquad (3.44b)$$

Again, since c can have either sign, we see that for one of the branches ω_o^2 is negative, hence there is an instability. Comparison with (3.40a) and (3.41c) shows that ω_o^2 has the opposite sign from the energy difference. This means that it is the upper branch which is unstable, while the lower branch shares the stability properites of the unique solution at the bifurcation point: if the latter is stable [the zero-eigenvalue mode is the lowest mode] so is the lower branch solution; if there is instability at the bifurcation point, [there exist complex eigenfrequencies] it will persist in the lower mode even beyond the bifurcation point.

When a similar derviation is carried out in the Yang-Mills problem, the conclusions are exactly the same [5]: one establishes that the static mode functions behave as $\pm(Q-Q_c)^{\frac{1}{2}}$ above the critical charge Q_c where bifurcation occurs. The energy difference between the two modes rises as $(Q-Q_c)^{\frac{3}{2}}$, with the upper branch being unstable. Since no analytic information about lower branch is available, we cannot report on its stability, but the general theory insures that it follows the stability properties at the bifurcation point. The fact that the zero-eigenvalue mode is spherically symmetric supports the conjecture that this is the lowest mode and no instability is present. Numerical analysis of the numerically determined results in Figs. 1 to 4 validate the above.

4. Conclusion

Finite-energy solutions to the Yang-Mills equations with arbitrary sources, can be studied perturbatively for weak sources. A rather comprehensive description is available. There exist at least two static solutions, the Abelian and non-Abelian Coulomb, with the latter carrying lower energy. They are accompanied by time-dependent generalizations which are continuous deformations of the static solutions. The time-dependent ones have the important property of lowering the energy, relative to the static configurations. The whole assembly of solutions can be compactly described in terms of the quantity $\vec{C}=[\Phi,\vec{\nabla}\Phi]$. Beyond the perturbative regime, it is difficult to study the problem analytically [save for the Abelian Coulomb case], but numerical investigation does not expose any significant new structure.

Some questions remain. One would like to know how many different non-Abelian Coulomb solutions there are for a fixed source. [Thus far we have found only one.] Also one wonders whether there is a topological distinction between the Abelian and non-Abelian cases; a hint of one arises from the observation that the gauge transformation U, which takes the source from the Abelian frame to the non-Abelian frame in (2.18), is topologically non-trivial.

Solutions which are supported only by sources that exceed a critical strength, are known in isolated examples, but little of a general nature can be said about them at present. Presumably they are always characterized by bifurcations, and one wonders whether the bifurcating solutions are topologically different from the perturbative ones. Again one finds a hint: at large distances, the radial non-Abelian Coulomb solution vanishes rapidly, while the bifurcating one tends to a non-trivial pure gauge. Also one would like to know how to characterize the different bifurcated branches.

Stability for weak sources can be established, but the behavior for stronger sources is thus far unknown, save for the Abelian Coulomb case where an explicit formula allows for computations -- the Coulomb solution is unstable beyond a critical source strength. In the bifurcating solutions, the bifurcation point corresponds to a zero-eigenvalue mode in the stability equations, and one of the bifurcated branches, the upper, is unstable. The other branch, the lower one, follows the stability behavior at the bifurcation point.

The most interesting result of the stability analysis for Yang-Mills theory is the observation that both the Abelian and non-Abelian Coulomb solutions, when stable, are gyroscopically stable. Modes which lower the energy, without introducing instability, have been identified. However, it is not clear how they are to be represented by superpositions of conventional monochromatic fluctuations.

The Yang-Mills model shares the physics of a top -- stable motion does not minimize the energy. The analogy can be developed. For the top, gyroscopic forces arise from the constraint of conservation of angular momentum. In the Yang-Mills theory, the gyroscopic terms arise by the imposition of the Gauss law constraint.

$$- \varepsilon_{abc}\vec{A}_b\cdot\vec{E}_c + \vec{\nabla}\cdot\vec{E}_a = \rho_a$$

But the left-hand side is the generator of local rotations in group space; it is like a group space angular momentum. [$- \varepsilon_{abc}\vec{A}_b\cdot\vec{E}_c$ is analogous $\vec{q} \times \vec{p}$.] In other words a non-vanishing source establishes at each point in ordinary space a non-vanishing angular momentum in group space, which then stabilizes configurations, which otherwise would be unstable [16].

While some further computations obviously suggest themselves, especially for strong sources, the most pressing open question concerns the relevance of these mathematical investigations to the quantum physics of Yang-Mills theory.

This lecture was presented at the Summer Workshop on Theoretical Physics, Trieste, Italy, 12 July - 3 August, 1979; at the Durham Symposium on Geometry and Physics, Durham, England, 11-21 July, 1979; at the Nuffield Workshop on Quantum Gravity, Cambridge, England, 20 July - 20 August, 1979; and at the CMS Summer Research Institute, Montreal, Canada, 3-8 September, 1979. The research was supported in part through funds provided by the U.S. Department of Energy under contract EY-76-C-02-3069.

References

1. R. Jackiw, L. Jacobs and C. Rebbi, Phys. Rev. D 20, 474 (1979).
2. P. Sikivie and N. Weiss, Phys. Rev. D 20, 487 (1979).
3. J. Mandula, Phys. Rev. D 14, 3497 (1976).
4. An example of this solution, with a radially symmetric source,

$\rho_a = \delta_{a3}q(r)$, is constructed by taking $\vec{A}_a = -\vec{E}_a t + \delta_{a1}\vec{\nabla}(\alpha r^2 \frac{d\phi}{dr})$,
$\vec{E}_a = \frac{-\hat{r}}{r^2}\frac{1}{\alpha}\{\delta_{a3}\sin(\alpha r^2\frac{d\phi}{dr}) + \delta_{a2}[1-\cos(\alpha r\frac{d\phi}{dr})]\}$, $\phi = -\frac{1}{\nabla^2}q$.
This is of the form (2.15) with $U = \exp(\frac{i\sigma^1}{2}\alpha r^2\frac{d\phi}{dr})$. The energy is $\mathcal{E} = 8\pi\int_0^\infty \frac{dr}{\alpha^2 r^2}\sin^2(\frac{\alpha}{2} r^2\frac{d\phi}{dr})$. Here α is an arbitrary parameter, which when set to zero gives the Abelian Coulomb solution. This configuration is essentially the "total screening solution" found by P. Sikivie and N. Weiss, Phys. Rev. Lett. 40, 1411 (1978), and Phys. Rev. D 18, 3809 (1978); except that they use a descrete parameter rather than our continuously varying α. The feasibility of generalizing the total screening solution was pointed out by P. Pirilä and P. Presnajder, Nucl. Phys. B 142, 229 (1978). The present formulation is given by Jackiw and Rossi, Ref. 5.

5. R. Jackiw and P. Rossi MIT preprint (to be published).

6. S. Coleman, in New Phenomena in Sub-Nuclear Physics, edited by A. Zichichi (Plenum, New York, 1977); S. Deser, Phys. Lett. 64B 463 (1976).

7. The truth of this statement is manifest from (2.22c) for sufficiently small Q, so that the $O(Q^4)$ terms are negligible; and for charge densities $q(r)$ which never change sign, so that $q(r)q(r')\hat{r}\cdot\hat{r}' < q(r)q(r')$. However, in Ref. 1 it is shown that even for charge densities with varying signs the inequality in (2.22c) is valid, and that numerical computation at large Q confirms the bound; see Fig. 1.

8. J. Goldstone (unpublished); see also Ref. 1.

9. Another example of a solution which exists when the source is of sufficient magnitude is Sikivie and Weiss' "magnetic dipole solution"; see Ref. 4. The source which supports this solution is studied by Y. Leroyer and A. Raychaudhuri, Phys. Rev. D (in press).

10. For an introduction to current, mainly mathermatical research on stability of motion see C. Siegel and J. Moser, Lectures on Celestial Mechanics, (Springer Verlag, Berlin, 1971). A simple discussion, referring to older physics research, is found in H. Jeffreys and B. Jeffreys, Methods of Mathematical Physics, (Cambridge University Press, Cambridge, 1972).

11. This form for the fluctuation equations in a Coulomb background field was also given by M. Magg, Physics Letters 74B, 246 (1978).

12. J. Mandula, Physics Letters 67B, 175 (1978); M. Magg, Physics Letters 74B, 246 (1978).

13. L. Schiff, H. Snyder and J. Weinberg, Phys. Rev. 57, 315 (1940); K. Johnson, Harvard Ph.D theisis (1954) (unpublished); A. Migdal, Zh. Eksp, Teor, Fiz. 61, 2209 (1972) [English translation: Soviet Physics JETP 34, 1184 (1972)]; A. Klein and J. Rafelski, Phys. Rev. D 11, 300 (1975). For the delta-shell source (2.34) the instability sets in at Q=1.5. This value is not related in any transparent way to the magnitude of Q at the bifurcation. It does agree with the strength of a point source at the onset of instability, as determined in Ref. 12. However this coincidence is a consequence of the scaling properties of the delta-shell source, and is not expected for arbitrary extended sources.

14. The truth of this statement is manifest for charge densities which do not change sign, so that $q(\vec{r})q(\vec{r}\,')>0$. However, for spherically symmetric charge densities the proviso can be removed, see Ref. 15, below.

15. For shperically symmetric charge densities, we may take the solution described in Ref. 4. For small α it becomes an infinitesimal deformation of the Abelian Coulomb with energy

$$\mathcal{E} = 2\pi \int_0^\infty dr\, r^2 (\phi')^2 - \frac{\alpha^2\pi}{6} \int_0^\infty dr\, r^6 (\phi')^4,$$ which is always less than

the Coulomb energy, regardless of the sign of the source.

16. That the physics of the top is encountered in the Yang-Mills theory was previously remarked by J. Goldstone and R. Jackiw, Phys. Lett. 74B, 81 (1978). Indeed it was in the context of the formalism developed in this paper that some of the results sumarized here were first encountered.

ON THE LONG-RANGE INTERACTION OF TOPOLOGICALLY CHARGED MONOPOLES

L. O'Raifeartaigh, Dublin Institute for Advanced Studies †

and

S.Y. Park and K.C. Wali, Syracuse University, New York

Introduction

The monopoles considered here will be the extended-source, finite-energy, solutions of the Yang-Mills-Higgs(YMH) equations with topological charge [1]. Since the monopoles are self-supporting and singularity-free, it is possible to compute their physical properties without encountering the external mechanical forces or point charges which are necessary to maintain the stability of the sources in linear theories. The physical properties of single monopoles are now fairly well-known and the forces between monopoles(especially the long-range forces) are one of the simplest properties of the multi-monopole system.

Actually, if all the physical(non-Goldstone) Higgs fields are massive, the long-range forces are not particularly interesting, since they are just the forces one would expect from a system of point magnetic charges. However, if some of the Higgs fields are massless they contribute a long-range scalar attraction, and in the case of like magnetic charges this attraction competes with the magnetostatic repulsion. The most extreme case occurs in the Prasad-Sommerfield(PS) limit [2], when the Higgs potential is set equal to zero after the spontaneous breakdown, so that <u>all</u> the Higgs fields are massless, and this is the case that we shall consider.

As one might expect, the long-range force between any two monopoles due to the massless gauge and Higgs fields is given by an inverse-square law of the form

$$F_{12} = \frac{k_1 k_2 - g_1 g_2}{|r_1 - r_2|^2} \quad , \tag{1.1}$$

where the k_s and g_s are the scalar and magnetic charges respectively. However, in contrast to the Maxwell-Newton systems of point charges and masses, the constants g_s and k_s are not assigned arbitrarily, but emerge as functionals of the single-monopole fields. The g_s are, of course, just the topological magnetic charges, and the k_s turn out to be proportional to the Higgs kinetic energy. One of the results presented in this paper is a general expression for k_s in terms of the fields (see (3.3)).

As an application of the formula (1.1) we use it to show that the Olive-Mentonen(OM) symmetry [3] between PS-monopoles and gauge fields can hold only if the Higgs fields belong to the adjoint representation of the gauge group. Details of the calculations sketched here can be found in [4].

2. Single-Monopoles

We recall that single monopoles are finite-energy solutions of the static-Yang-Mills-Higgs system with Hamiltonian

$$H = \int \tfrac{1}{2}(F,F) + \tfrac{1}{2}(D\Phi,D\Phi) + V(\Phi), \tag{2.1}$$

† Presented by L. O'Raifeartaigh.

where

$$F = \nabla \times A + e[A,A] \quad , \quad D\Phi = \nabla\Phi + A\wedge\Phi, \tag{2.2}$$

$\vec{A}$ is the gauge-potential for any compact Lie group G, and Φ are scalar(Higgs) fields belonging to any representation R of G. We shall assume that R is a real representation. The bracket denotes inner product in the relevant representation space, and the minimum of the potential occurs for $(\Phi,\Phi) = c^2$, where c is a non-zero constant. The field equations are

$$D^2\Phi = V'(\Phi) \;, \quad D.F = 0 \;, \quad D \times F = J = (t\Phi, D\Phi) \;, \tag{2.3}$$

where t are the group generators and the centre equation is just the Bianchi identity. It is now well-known [5] **that the finite energy solutions can be characterized by the** homotopy class of $\phi(\Omega)$, where (r,Ω) are polar coordinates and $\Phi(x) \to c\;\phi(\Omega)$ as $r\to\infty$, and that if the stability group of $\phi(\Omega)$ is u(1) the homotopy class is determined by an integer n. In that case the solution is identified as a magnetic monopole of charge $g = ne^{-1}$, and the charge g can always be expressed as a functional of $\phi(\Omega)$. For example, for $\phi(\Omega)$ in the adjoint representation of SU(2) one has

$$g = \frac{1}{2e}\int d\vec{\Omega}\;(\phi,\vec{L}\phi\wedge\vec{L}\phi) \;, \tag{2.4}$$

where $\vec{L}$ is the angular momentum operator and wedge denotes outer product in both group and configuration space. A rigorous mathematical proof [6] of the existence of finite-energy solutions to (2.1) exists only in the spherically symmetric SU(2)-case(for which n=1) but the topological considerations make it likely that solutions exist for other groups and other spatial configurations. Hence we shall make no restrictions regarding groups or configurations, except to assume that when the Higgs field is not in the adjoint representation, its stability group is u(1) i.e. the photon is the only massless gauge field.

3. The PS-limit and the Asymptotic Higgs Field

In the PS-limit the potential V in (2.1) is set equal to zero and replaced by the boundary condition $\Phi(x)\to c\phi(\Omega)$ as $r \to \infty$. Since the Higgs fields are then long-range their asymptotic form will be

$$\Phi(x) \to c\phi(\Omega) + \frac{k}{r}\,\psi(\Omega), \tag{3.1}$$

where ϕ and ψ are normalized to unity and k is a constant(the scalar charge). The finiteness of the energy requires that ϕ be covariantly constant and the field equations then require that ψ be covariantly constant and orthogonal to ϕ with respect to the generators [4]. Thus we have

$$D\psi = D\phi = 0 \;, \qquad (\phi,\, t\psi) = 0 \;, \tag{3.2}$$

The t-orthogonality essentially says that ψ is physical(orthogonal to the Goldstone directions $t\phi$). In particular, the conditions (3.2) allow ψ to be identical to ϕ, which is the case for spherical symmetric configurations. Since in the PS-limit V=0 the first field equation in (2.3) reduces to $D^2\Phi = 0$, the constant k can be expressed as a functional of the fields by partially integrating the Higgs kinetic energy. One finds, using (3.1) and (3.2) that

$$\int d^2x(D\Phi)^2 = \int d^3x\nabla(\Phi,D\Phi) = kc\int d\Omega(\phi,\psi) \quad , \tag{3.3}$$

which is the required expression. As indicated by the space-integral in (3.3) the scalar charge k is a global constant whose value is determined by the boundary conditions at the origin as well as infinity.

4. The PS-limit and Bogomolny Bound

A particularly interesting case of the PS-limit is when the Higgs field belongs to the adjoint representation of the gauge-group. The the Hamiltonian (2.1) can be written as

$$H = \tfrac{1}{2}\int d^3x \{F^2 + (D\Phi)^2\} = \tfrac{1}{2}\int d^3x(F\pm D\Phi)^2 + \tfrac{1}{2}g \quad , \tag{4.1}$$

where g is a topological charge. (For SU(2) g is the unique topological charge possible and is just the magnetic topological charge (2.4)) It follows from (3.1) that the energy, or mass, of a finite-energy solutions with topological charge g is bounded below by g/2. The bound g/2 is called the Bogomolny bound [7] and it is clearly saturated, if, and only if,

$$D\Phi = \pm F \tag{4.2}$$

Eq. (4.2) is a first-order differential equation whose solutions automatically satisfy the field equations (2.3). The solution in the spherically symmetric SU(2) case has been found explicitly [2] in terms of elementary hyperbolic functions. More general solutions (with $g = ne^{-1}$ where $n > 1$) would correspond to static n-monopole or n-anti-monopole (but not mixed monopole-anti-monopole) configurations. So far no such solutions have been found or even shown rigorously to exist.

5. The Long-Range Forces

In order to see how the long-range YMH forces should be calculated, let us first recall the Maxwell-Newton procedure for point sources. The MN procedure is to solve the static field equations

$$\nabla^2\phi = 0 \, , \qquad \nabla . f = 0 \, , \qquad \nabla\times f = 0 \, , \tag{5.1}$$

in a region E exterior to the sources, with boundary conditions corresponding to sources of charge g_s and mass k_s at points $\vec{r}_s$, to obtain

$$\phi = \Sigma_s \frac{k_s}{(r-r_s)} \quad , \qquad \vec{f} = \vec{\nabla} \sum_s \frac{g_s}{(r-r_s)} \quad . \tag{5.2}$$

Inserting (6.2) into the energy momentum tensor gives the usual inverse-square law(1.1).

To proceed analogously in the YMH case one must first take account of the finite size of the monopoles by letting the typical distance d between them be much larger than their cores ρ , and then defining the exterior region E to be all space outside a set of spheres of radius a surrounding the monopoles ($d \gg a \gg \rho$). Then the analogy to the MN procedure is to seek exact solutions of the static YMH equations (2.3)in the exterior region E, with J = 0,and with boundary conditions on the surface of each a-sphere which correspond to the asymptotic form of the single monopole solution inside the sphere.

6. Ansatz for Exact Exterior Solutions

To obtain an Ansatz for an exact exterior solution we first consider the case when $G = Su(2)$ and Φ is in the adjoint representation, because in that case the exterior condition $J = 0$ forces a unique Ansatz. In fact from the definition of $\vec{J}$ in (2.3) we have

$$\vec{J} = 0 \rightarrow \vec{D}\,\Phi \perp t\,\Phi \rightarrow \vec{D}\,\Phi \,||\, \Phi \tag{6.1}$$

where the orthogonal and parallel signs refer to the group space. Note that the last relation follows automatically only for Φ in the adjoint of Su(2). If we then write Φ as

$$\Phi = h(x)\phi(x), \tag{6.2}$$

where h is the norm and ϕ is normalized to unity, we see at once that

$$D\phi = 0 \tag{6.3}$$

Thus for Φ in the adjoint of Su(2) $\vec{J} = 0$ implies that ϕ is covariantly constant, and conversely.

This result together with the asymptotic form (3.1) on the surface of each a-sphere suggest the following Ansatz for Φ in the exterior region E

$$\Phi(x) = c\phi(x) + h(x)\psi(x) \quad \text{where } (\psi, t\phi) = 0 , \tag{6.4}$$

and the normalized fields $\phi(x)$ and $\psi(x)$ are covariantly constant. It is trivial to verify that this Ansatz implies $J = 0$ for all groupsG and representations R of Φ.

7. Exact Exterior Solutions

The Ansatz (6.4) does much more than annihilate the current $\vec{J}$ in E. It reduces the non-linear, non-abelian, YMH system, to the linear, abelian, MN system and hence allows us to solve the YMH system in terms of the known solutions (5.2) of the MN system. The key point in the reduction is to note that the integrability condition for the covariant constancy of ϕ is

$$[\vec{D}, \vec{D}]\,\phi = \vec{F}\,\phi = 0 , \tag{7.1}$$

and since the stability group of $\phi(x)$ is assumed to be U(1) this shows that $\vec{F}$ must factorize into

$$\vec{F} = n(x)\vec{f}(x) \quad \text{where } (n,n) = 1 \text{ and } n\phi = 0 , \tag{7.2}$$

where n(x) is the electromagnetic direction in the Lie algebra. It is easily verified that if the stability group of $\phi(x)$ is U(1), n(x) is covariantly constant and that by inserting (6.4) and (7.2) into the YMH system, it reduces to the MN system (5.2) for f and h . The only further condition is the integrability condition for the covariant constancy of $\psi(x)$, which reduces to $n(x)\psi(x) = 0$. Thus an exact exterior solution for the YMH system is

$$\Phi(x) = c\phi(x) + \psi(x) \sum_s \frac{k_s}{|r-r_s|} , \tag{7.3}$$

$$\vec{F}(x) = n(x)\,\vec{\nabla}\,\Sigma_s \frac{g_s}{|r-r_s|} \,, \tag{7.4}$$

where ϕ,ψ and n are normalized and covariantly constant and satisfy the algebraic conditions

$$n\phi = n\psi = 0 \,, \qquad (\psi, t\phi) = 0 \,. \tag{7.5}$$

Note that the procedure is to first choose a $\phi(x)$ with the right topological properties to describe a monopole of strength g_s on the surface of each a-sphere(this defines $\phi(x)$ up to a gauge transformation) and then to determine A(x), n(x) and $\psi(x)$ from the covariant constancy of ϕ and from the field equations.

From (7.3), (7.4) it is clear that the long-range forces are just the MN inverse-square law forces mentioned in (1.1), and the constants g_s and k_s are easily identified with the magnetic and scalar charges of the individual monopoles, as discussed in sect.3. In particular once g_s is given for each monopole the k_s can be determined in principle from the field equations. We shall now consider the determination of k_s for two special cases:
(i) When the Bogomolny bound is saturated.
(ii) When G=SU(2), the solution is spherically symmetric, and Φ belongs to an arbitrary integer spin representation.

8. The Long-Range Forces and the Bogomolny Bound

From (4.1) and (4.2) we see that if the Bogomolny Bound is saturated, there is an equipartition of energy between the pure gauge field F^2 and the kinetic contribution of the Higgs field $(D\phi)^2$. In fact,we have

$$H = 2\int d^2x (D\Phi)^2 = 2\int d^3x\, F^2 = 2g \,. \tag{8.1}$$

But from the expression (3.3), we see that the scalar constant then reduces to

$$k = g \,/\, (\phi,\psi) \;\geq\; g, \tag{8.2}$$

where there is equality if, and only if, $\phi = \psi$. Thus in general the scalar forces dominate. However if $\phi = \psi$, which is true in the spherically symmetric case at least, the scalar and magnetostatic forces are just equal in magnitude. In particular, for like spherically symmetric magnetic charges, the magnetostatic repulsion is just balanced by the scalar attraction and there is no long-range force. The vanishing of the long-range force for spherically symmetric monopoles was first pointed out by Manton [8] for the SU(2) gauge group.

9. Spherically Symmetric SU(2) Monopoles for $I \neq 1$

The case of spherically symmetric SU(2) monopoles when the Higgs field is not in the adjoint representation has been studied in [9]. In order to have a U(1)-stability-group the Higgs field must lie on a special orbit in an integer spin representation with Casimir invariant $s = I(I+1)/2$ and on making the spherically symmetric Ansatz $\Phi(x) = \phi(\Omega)h(r)$ and $A(K) = a(\Omega)\,(K(r)-1)/r$ the Hamiltonian (2.1) reduces to

$$H(s) = \frac{4\pi}{e^2}\int_0^\infty dr\,\{\,(K')^2 + \tfrac{1}{2}\,\frac{K^2-1}{2} \;+\frac{r^2}{2}\,(h')^2 + sK^2h^2\}. \tag{9.1}$$

Solutions to the field equations for this Hamiltonian have been shown to exist, and corresponding energy H(s) has been shown to be a monotonically increasing function of s, which is bounded above as $s\to\infty$.

Because of the spherical symmetry [10] the magnetic charge g is unity for all s, and it is easy to see that the scalar charge k reduces to

$$k(s) = \frac{4\pi}{e^2 c} \int_0^\infty dr \{ \frac{r^2}{2}(h')^2 + sK^2h^2\} \qquad (9.2)$$

Since the explicit solutions of the field equations are not known for $s \neq 1$ it is not possible to evaluate k(s). But since $ck(s) < H(s)$ it follows that k(s) is bounded as $s \to \infty$. Furthermore (9.2) suggests that k(s) is not independent of s, but increases monotonically, and an estimate of the limiting case $s = \infty$ which yields $k(\infty)/k(1) \simeq 3$ supports this view. The cancellation of the long-range forces would require $k(s) = g$ for all s, that is, it would require k(s) to be constant, and hence such a cancellation would seem to be highly unlikely. In particular, if k(s) increases monotonically with s, then the scalar forces dominate for $s > 1$.

10. Monopole Gauge-Field Symmetries

Olive and Mentonen [3] have postulated a dual relationship between the monopoles and the massive gauge-fields from which they are constructed, and some support for this duality comes from the following two remarkable symmetries between the monopoles and gauge fields, which occur when the Bogomolny bound is saturated for spherically symmetric SU(2) monopoles.

(i) The gauge-field and monopole masses are $M_W = ec$ and $M_m = gc$, respectively where e and g are the Yang-Mills and topological charges. This result is interesting because M_W and M_m are obtained by very different means, namely M_W by the Higgs mechanism and M_m by a dynamical calculation of the energy.

(ii) For both the monopoles and massive gauge-fields the long range forces cancel. Again the result is obtained in a different manner in each case, namely from the calculation of sect. 9 for the monopoles and from the Born graphs(one photon plus one scalar exchange) for the gauge-fields.

The question is whether the gauge-field-monopole symmetry persists when the Higgs field is no longer in the adjoint representation, and using the results of sect. 9 one can show at once that it does not.

First, it is easy to see that for isospin I the Higgs mechanism generates the gauge-field mass $M_W = ecs^{\frac{1}{2}}$, where $s = I(I+1)|2$. On the other hand, the mass of the monopole is just the energy H(s), and as mentioned in sect. 9, H(s) is bounded above as $s \to \infty$. Thus the s-dependence of M_W and M_m is quite different.

Second, even when the Higgs field has isospin I, the gauge field remains in the adjoint representation, and hence for $I \neq 1$ the coupling constants for the photon and scalar vertices of the Born graphs become eM_W and e^2cs respectively. Thus the net long-range force for the gauge-fields is proportional to $e^4c^2s(s-1) \mid r^2$. On the other hand, from the results of sect. 10, the net long-range force for the monopoles becomes $(k^2(s)-g^2)|r^2$ where k(s) is bounded above as $s \to \infty$. It follows that the two net forces cannot have the same s-dependence.

Thus for both the masses and long-range forces the gauge-field-monopole symmetry breaks down when the Higgs field is not in the adjoint representation.

References

1. P. Goddard and D. Olive,(Review) Reports on Progress in Physics 41, 1357 (1978)
2. M. Prasad and C. Sommerfield, Phys. Rev. Letters 35 760 (1975)
3. D. Olive and C. Montonen. Phys. Lett. 72B, 117 (1977)
4. L. O'Raifeartaigh, S.Y. Park and K.C. Wali Phys. Rev. (in press)
5. M. Monastyrskii and A. Perelmov JETP Letters 21, 43 (1975)
   J. Arafune, P. Freund and C. Goebel. J.Math.Phys. 16, 433 (1975)
6. Y. Tyupkin, V. Fateev and A. Shvarts, Teor.Mat.Fiz. 26(1976)270
7. E. Bogomolny, Sov.J.Nucl.Phys. 24, 449 (1976)
8. N. Manton, Nucl. Phys. B126, 525 (1977)
   J. Jersak, M. Kiera and M. Magg, Nuovo Cim. 40A, 269 (1977)
9. L. O'Raifeartaigh and J. Rawnsley, Phys. Letters 72B, 465 (1978)
10. A. Guth and E. Weinberg, Phys. Rev. D14, 1660 (1976)
   L. O'Raifeartaigh, Nuovo Cim. Letters 18, 205 (1976)

INTERACTION OF SUPERCONDUCTING VORTICES

Lectures presented at the Canadian Mathematical Society
Summer Research Institute on Gauge Theories

Montreal, Canada September 1979

Claudio Rebbi
Brookhaven National Laboratory
Department of Physics
Upton, New York 11973

I. Introduction

An Abelian gauge field minimally coupled to a charged matter field in two dimensions constitutes one of the simplest examples of a system with interesting topological properties. The model is physically relevant: it can be used, as in the Ginzburg-Landau theory, to describe cross-sections of superconductors with translational symmetry along some axis [1] ; more recently it has been applied to the physics of elementary particles [2] .

A topological quantum number q, related to the boundary values of the fields, characterizes all finite energy configurations. In the sector with $|q| = 1$ localized vortex-like solutions of the field equations are known to exist [3] . Multi-vortex solutions with $|q| > 1$ exist in special cases [4] . In general, configurations with many vortices are not in equilibrium.

In these two lectures I shall describe results recently obtained on multi-vortex configurations. After a brief review of the model, I shall first illustrate a numerical analysis, performed by variational methods, of the interaction between two vortices [5] . The study, done in collaboration with Laurence Jacobs, shows that two vortices attract or repel each other according to whether a dimensionless coupling constant λ, characterizing the relative strength of the matter self-coupling versus the gauge coupling, takes a value smaller or greater than one. (This agrees with results previously obtained for asymptotic separations of the vortices [6]). For $\lambda = 1$, in particular, the vortices appear in equilibrium at any separation, hinting to the existence of a much wider class of solutions to the field equations. In the second lecture I shall consider in detail the case $\lambda = 1$, illustrating analytical results which demonstrate that for this special value of the coupling constant solutions with any number of vortices at arbitrary positions do indeed exist [5,7,8, 9] .

II. Interaction Energy of Two Vortices

The model is defined by the following energy functional:

$$E = \int d^2x\left[\tfrac{1}{2}|(\partial_i - ieA_i)\phi|^2 + \tfrac{1}{4}F_{ij}F^{ij} + c_4(|\phi|^2 - c_0^{\,2})^2\right]. \qquad (2.1)$$

A_i stands for the two components of an Abelian gauge potential; $F_{ij} = \partial_i A_j - \partial_j A_i = \varepsilon_{ij}B$, $B = \tfrac{1}{2}\varepsilon_{lm}F^{lm}$; ϕ is a complex matter field and the integration is over the two-dimensional plane. In the application to superconductivity ϕ is the "gap parameter", describing the density of superconducting pairs, B is the magnitude of the magnetic field oriented orthogonally to the plane and E is the free energy per unit length. In the applications to particle physics ϕ represents a Higgs field, since its vacuum value is different from zero, $|\phi_{\text{vacuum}}| = c_0$, and is coupled to a gauge potential.

Some of the coupling constants can be absorbed into a rescaling of the fields. Defining

$$x^i = \frac{1}{ec_0}\tilde{x}^i, \quad \phi = c_0\tilde{\phi}, \quad A_i = c_0\tilde{A}_i$$

the energy functional takes the form

$$E = \frac{c_0}{e}\int d^2x\left[\tfrac{1}{2}|(\tilde{\partial}_i - i\tilde{A}_i)\tilde{\phi}|^2 + \tfrac{1}{4}\tilde{F}_{ij}\tilde{F}^{ij} + 1/8\,\lambda^2(|\tilde{\phi}|^2 - 1)^2\right] \tag{2.2}$$

$$\text{with } \lambda^2 = \frac{8c_4}{e^2}. \tag{2.3}$$

λ, which measures the relative strengths of the self-coupling of the matter field versus the gauge coupling, is a physically relevant parameter.

It shall prove quite convenient to use a complex notation for the coordinates in the plane and the gauge potentials. We define therefore

$$z = \tilde{x}_1 + i\tilde{x}_2, \quad \bar{z} = \tilde{x}_1 - i\tilde{x}_2, \tag{2.4a}$$

$$\partial \equiv \frac{\partial}{\partial z} = \tfrac{1}{2}\left(\frac{\partial}{\partial\tilde{x}_1} - i\frac{\partial}{\partial\tilde{x}_2}\right), \quad \bar{\partial} \equiv \frac{\partial}{\partial\bar{z}} = \tfrac{1}{2}\left(\frac{\partial}{\partial\tilde{x}_1} + \frac{i\partial}{\partial\tilde{x}_2}\right), \tag{2.4b}$$

$$A = \tfrac{1}{2}(\tilde{A}_1 - i\tilde{A}_2), \quad \bar{A} = \tfrac{1}{2}(\tilde{A}_1 + i\tilde{A}_2). \tag{2.4c}$$

In terms of these variables the energy functional is

$$E = \frac{c_0\pi}{e}\mathcal{E}, \tag{2.5}$$

with

$$\mathcal{E} = \frac{1}{2\pi}\int dz d\bar{z}\,\left[|(\partial - iA)\tilde{\phi}|^2 + |(\bar{\partial} - i\bar{A})\tilde{\phi}|^2 + 2|\bar{\partial}A - \partial\bar{A}|^2 + \right.$$
$$\left. + 1/8\,\lambda^2(|\tilde{\phi}|^2 - 1)^2\right] \tag{2.6}$$

The tilde over ϕ will be omitted from now on.

Finiteness of the energy demands

$$\lim_{|z|\to\infty} |\phi| = 1 \tag{2.7}$$

and

$$\lim_{|z|\to\infty} (\partial - iA)\phi = 0. \tag{2.8}$$

From (2.7) we see that ϕ must approach the value $e^{iX(\theta)}$ as $z \to \infty$ with fixed argument θ. Also, from continuity,

$$X(\theta + 2\pi) = X(\theta) + 2\pi q, \tag{2.9}$$

where q is an integer. Eq. (2.8) implies that, as $z \to \infty$, A must tend to a pure gauge form,

$$A = -i\partial \ln \phi + O\left(\frac{1}{|z|}\right) = X + O\left(\frac{1}{|z|}\right). \tag{2.10}$$

Gauss' theorem then gives the total magnetic flux through the plane:

$$\phi(B) = -\frac{i}{e}\int dz d\bar{z}(\partial\bar{A} - \bar{\partial}A) = \frac{1}{e}\lim_{|z|\to\infty}\oint(Adz + \bar{A}d\bar{z}) =$$

$$= \frac{1}{e}\oint dX = \frac{2\pi q}{e}. \tag{2.11}$$

The field configurations with finite energy are therefore divided into classes, labelled by q. Each class contains all those finite-energy field configurations which can be continuously distorted into each other (hence the denomination of topological quantum number for q) and within each class the total magnetic flux is $\frac{2\pi q}{e}$.

Notice that q is also given by

$$q = \frac{-i}{2\pi}\oint_\gamma d\ln\phi, \tag{2.12}$$

where γ is a closed contour enclosing all zeroes of ϕ. Distorting γ with continuity the integral can change only when the contour crosses a zero of ϕ. It follows $q = n_+ - n_-$, n_+, n_- being the numbers of zeroes of ϕ where ϕ vanishes as $z - z_i^{(+)}$ or $\bar{z} - \bar{z}_i^{(-)}$, respectively. We shall say that the field configuration exhibits vortices at the points $z_i^{(+)}$ and anti-vortices at the points $z_i^{(-)}$. Then q is the difference between the number of vortices and the number of anti-vortices.

Fields which make the energy stationary obey the Euler-Lagrange equations

$$(\partial - iA)(\bar{\partial} - i\bar{A})\phi + (\bar{\partial} - i\bar{A})(\partial - iA)\phi - \tfrac{1}{4}\lambda^2(\phi\bar{\phi} - 1) = 0, \tag{2.13}$$

$$4\partial\bar{\partial}A - 4\partial^2 A - i\bar{\phi}\partial\phi + i\phi\partial\bar{\phi} - 2A\phi\bar{\phi} = 0. \tag{2.14}$$

These equations are invariant under rotations of the plane. It is therefore possible to satisfy them with a rotationally symmetric Ansatz of the form

$$\phi = e^{iq\theta} f(r),$$

$$A = -(qi/2z)a(r),$$

$$r = |z|, \; f(\infty) = a(\infty) = 1. \tag{2.15}$$

The boundary behavior of the fields characterizes them as having vorticity q. It follows in particular that ϕ must have q zeroes, which, because of the symmetry, must be at the origin. Hence $f(r) = O(r^q)$ for $r \to 0$. Regularity of $A(z,\bar{z})$ demands $a(r) = O(r^2)$ for $r \to 0$.

Inserting the Ansatz of Eq. (2.15) into the Euler-Lagrange equations, these reduce to

$$\frac{d^2f}{dr^2} + \frac{1}{r}\frac{df}{dr} - \frac{q^2(a-1)^2}{r^2} - \frac{1}{2}\lambda^2 f(f^2 - 1) = 0, \tag{2.16}$$

$$\frac{d^2a}{dr^2} - \frac{1}{r}\frac{da}{dr} - (a-1)f^2 = 0. \tag{2.17}$$

Equations (2.16) and (2.17) cannot be solved analytically, but it is straightforward to verify that the values f = a = 1 are approached exponentially as

$$f(r) - 1 = O(e^{-\lambda r}),$$

$$a(r) - 1 = O(e^{-r}) \tag{2.18}$$

for $r \to \infty$. The coupling constant λ thus specifies the relative rate of spatial decay of the matter field versus that of the electromagnetic field. For intermediate values of r the equations must be solved numerically. Alternatively, an Ansatz, compatible with Eqs. (2.15) and (2.18) and containing further variational parameters, may be inserted into the expression for the energy, which is then minimized. Such a computation has been carried out for q = 1 and 2 by Jacobs and Rebbi [5] and I shall describe now briefly the method followed and the results.

The variational Ansatz is obtained by further specializing the functions f(r) and a(r) through an expansion

$$f(r) = 1 + e^{-\lambda r} \sum_{n=0}^{N} (f_n r^n/n!), \tag{2.19}$$

$$a(r) = 1 + e^{-r} \sum_{n=0}^{N} (a_n r^n/n!), \tag{2.20}$$

which, in the actual computation, has been truncated at N = 10. f_0, a_0, a_1 are set equal to -1 and f_1 is set equal to $-\lambda$ for vorticity two in order to reproduce the correct behavior at the origin of the fields. The remaining coefficients of the expansion are the variational parameters.

Inserting the Ansatz into the expression for $\mathcal{E}$ produces a polynomial function of the f_n and a_n variables, which will be denoted collectively by v_n. The polynomial is of the fourth order and all the coefficients can be evaluated analytically. A convenient numerical procedure to find a minimum consists in approximating the quartic $\mathcal{E}(v_n)$ with a quadric, tangent to it (or more precisely osculating it) at a definite point $v_n^{(0)}$ The minimum $v_n^{(1)}$ of this quadric is found and taken as an approximation to the true minimum of $\mathcal{E}$. The energy surface is then approximated again by a quadric, but now tangent to it at $v_n^{(1)}$ and the whole procedure iterated. The method converges very fast, undoubtedly also because of good convexity properties of the energy functional, depending on the physical nature of the problem. The values we found for the energies of a single vortex and of two superimposed vortices are reproduced in the following table and in Fig. 1.

λ	$E(\lambda,\ =1)$	$E(\lambda,\ =2)$	$\Delta E=E(\ =2)-2E(\ =1)$
0.5	0.75742	1.39129	-0.12355
0.6	0.81305	1.52627	-0.09983
0.7	0.86440	1.65337	-0.07543
0.8	0.91230	1.77407	-0.05053
0.9	0.95736	1.88936	-0.02536
1.0	1.00000	2.00000	0.00000
1.1	1.04053	2.10655	0.02549
1.2	1.07921	2.20945	0.05103
1.3	1.11625	2.30905	0.07655
1.4	1.15181	2.40567	0.10205
1.5	1.18639	2.49953	0.12675

It is apparent from these numerical results that for $\lambda < 1$ a configuration of two superimposed vortices has a lower energy than one of two vortices asymptotically separated, whereas the opposite is true for $\lambda > 1$. It becomes interesting then to follow the behavior of the energy as a function of the separation of the vortices. This cannot be determined from a solution of the Euler-Lagrange equations, which will not be satisfied unless the energy is at a stationary point, but can be obtained from a variational minimization of $\mathcal{E}$ subject to the constraint that ϕ have two zeroes (vortices) at a fixed separation.

A two-vortex Ansatz is constructed as follows. Let us denote by $f^{(q)}(r)$ and $a^{(q)}(r)$ $(q = 1,2)$ the functions determined numerically in the variational search for solutions with rotational symmetry. We set then

$$\phi(z,\bar{z}) = \left\{ \left[z^2 - \left(\frac{d}{2}\right)^2\right] / \left[\bar{z}^2 - \left(\frac{d}{2}\right)^2\right] \right\}^{\frac{1}{2}} f(z,\bar{z}), \tag{2.21}$$

$$f(z,\bar{z}) = w f^{(1)}\left[\left|z-\frac{d}{2}\right|\right] f^{(1)}\left[\left|z+\frac{d}{2}\right|\right] + $$

$$+ (1-w) \frac{\left|z^2 - (d/2)^2\right|}{|z|^2} f^{(2)}(|z|) + \delta f(z,\bar{z}), \tag{2.22}$$

$$A = w\left\{ -\frac{i}{2z-d}\, a^{(1)}\left[\left|z - \frac{d}{2}\right|\right] - \frac{i}{2z+d}\, a^{(1)}\left[\left|z + \frac{d}{2}\right|\right]\right\}$$

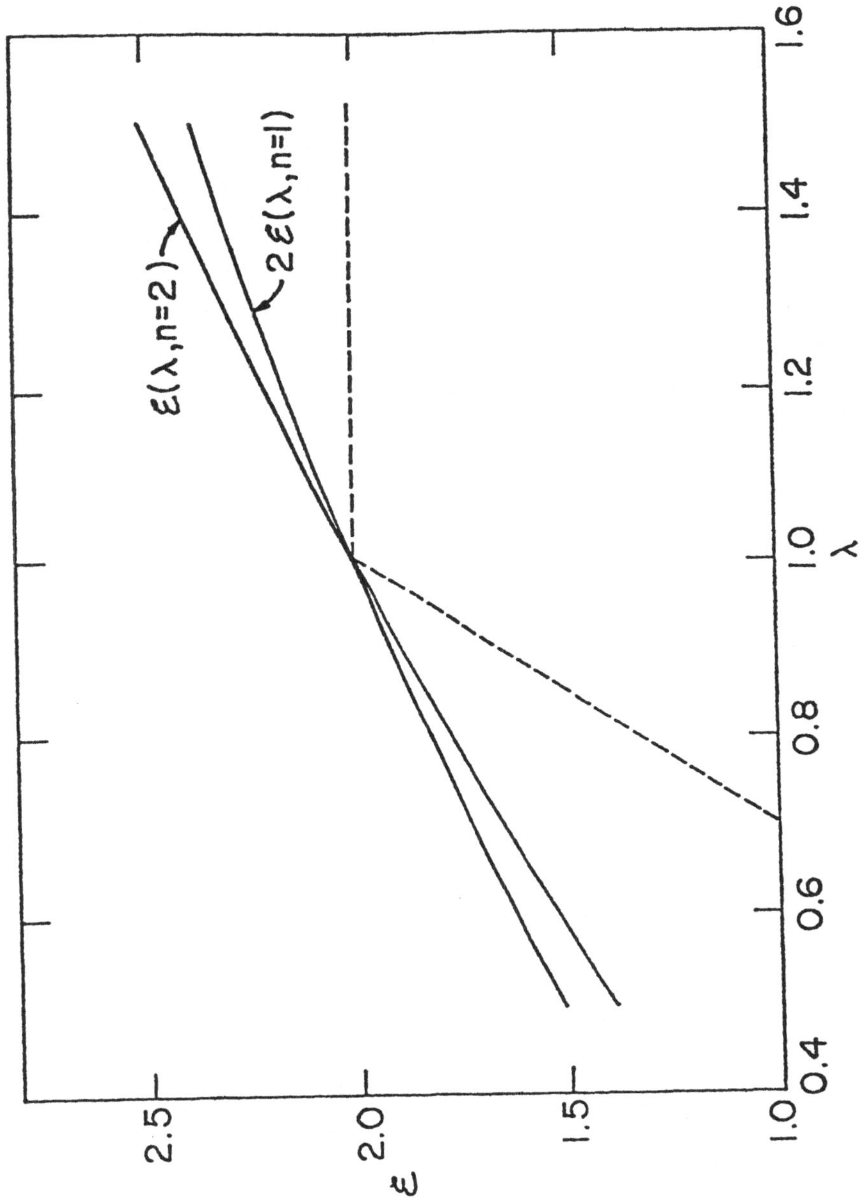

Fig. 1

Energies of two superimposed octationally symmetric vortices vs. two asymptotically repeated ones for various coupling strength ratios λ .

$$- (1 - w)\frac{i}{z}\, a^{(2)}(|z|) + \delta a(z,\bar{z}). \qquad (2.23)$$

An explanation of the various terms in this Ansatz follows. The first factor in the r.h.s. of Eq. (2.21) defines the phase of the matter field ϕ. The phase changes by 4π along any circle enclosing the points $z = \pm\frac{d}{2}$ so that $q = 2$. In a small loop around $z = \frac{d}{2}$ or $z = -\frac{d}{2}$ the phase changes by 2π; these are then the location of the vortices and f must vanish there. Notice that expressing ϕ in terms of a fixed phase factor times a real function f is equivalent to a choice of gauge (a generalized unitary gauge). The first terms in the r.h.s. of Eqs. (2.22) and (2.23) (excluding δf and δa) represent a zero order Ansatz, in which two single vortex configurations centered at $z = \frac{d}{2}$ and $z = -\frac{d}{2}$ are mixed with a symmetric two vortex configuration centered at the origin. A few factors are inserted to guarantee the appropriate behavior of the fields asymptotically and at the vortices; w is a mixing parameter, which is in a sense also a variational parameter, but in the actual computation it has been chosen minimizing $\mathcal{E}$ with $\delta f = \delta a = 0$; w approaches zero as $d \to 0$ whereas it approaches 1 as $\frac{d}{2}$ increases beyond the range of the vortices.

$\delta f(z,\bar{z})$ and $\delta a(z,\bar{z})$ are the variational terms proper. These functions have been expanded as

$$\delta f(z,\bar{z}) = \left|z^2 - \left(\frac{d}{2}\right)^2\right| (\cos\lambda|z|)^{-1} \times$$

$$\times \sum_{n=0}^{N} \sum_{m=0}^{n} f_{mn} \frac{(z\bar{z})^n}{2} \left[\left(\frac{z}{\bar{z}}\right)^m + \left(\frac{\bar{z}}{z}\right)^m \right], \qquad (2.24)$$

$$\delta a(z,\bar{z}) = \frac{1}{\cosh|z|} \left[z a^{I}(z,\bar{z}) + \bar{z} a^{II}(z,\bar{z}) \right] \qquad (2.25)$$

$$a^{I(II)} = \sum_{n=0}^{N} \sum_{m=0}^{n} a^{I(II)}_{mn} \frac{(z\bar{z})^n}{2} \left[\left(\frac{z}{\bar{z}}\right)^m + \left(\frac{\bar{z}}{z}\right)^m \right] \qquad (2.26)$$

The coefficients f_{mn}, a^{I}_{mn} and a^{II}_{mn} are the variational parameters. Eqs. (2.24) $\div$ (2.26), apart from a few factors inserted to reproduce the correct zeroes and asymptotic behavior of the fields, represent expansions into a series of terms $r^{2n}\cos 2m\theta$, compatible with the symmetry of the configuration. In the numerical computations N was set equal to 2, giving a total of 18 variational parameters. A few checks were performed with a larger value for N.

As in the case with rotational symmetry, the energy functional $\mathcal{E}$ is a quartic polynomial in the variational parameters. Once the coefficients of this polynomial are evaluated (because of the more complex configuration of the fields some coefficients must now be determined by numerical integrations), the iterative procedure already described for the rotationally symmetric configuration can be employed again very efficiently and the constrained minimum of $\mathcal{E}$ is found.

The results obtained for the energy of two interacting vortices, for three representative values of λ and a few values of the separation d, are presented in the following table and in Figure 2.

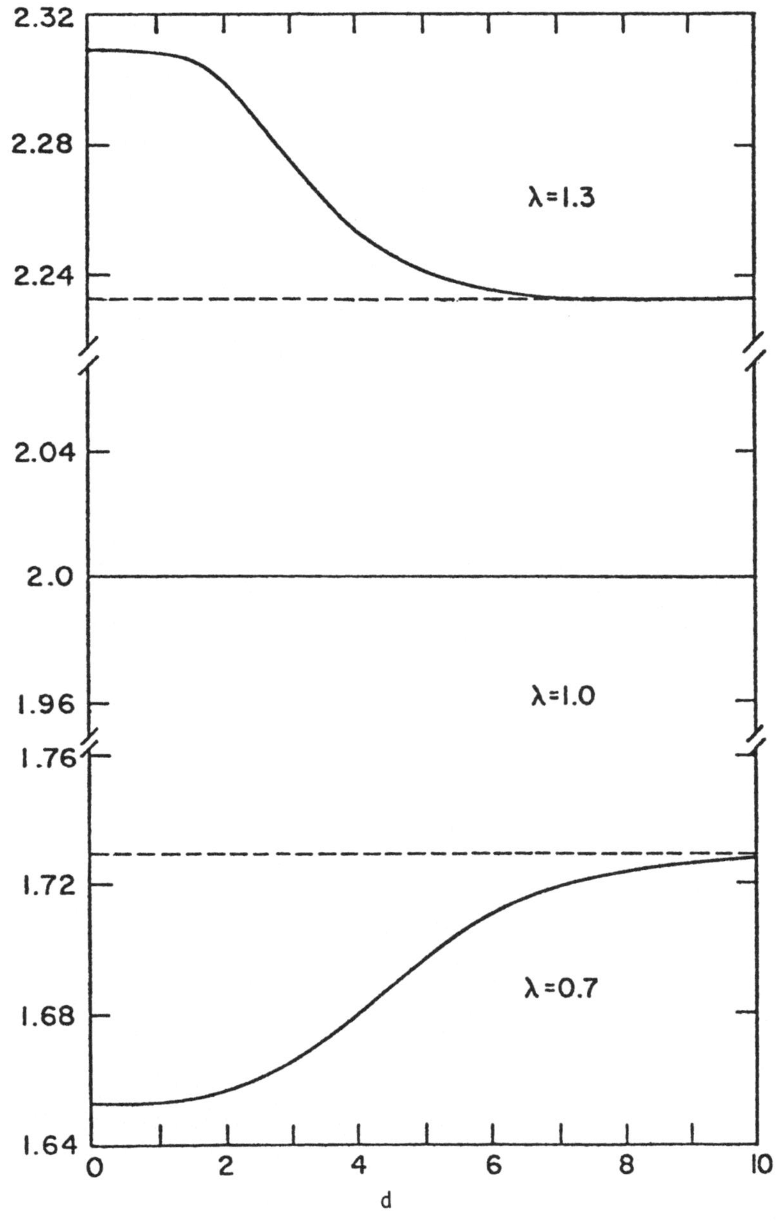

Fig. 2

Energy of two interacting vortices vs. separation distance, for $\lambda = 1.3$, 1.0 and 0.7.

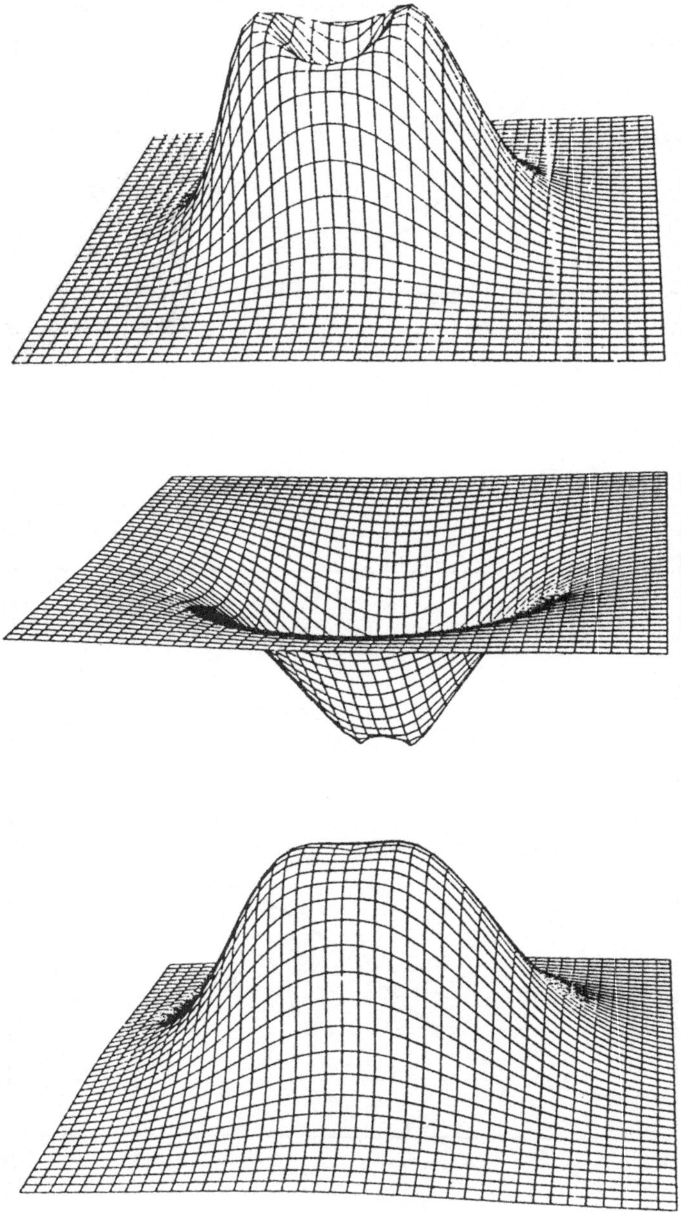

Fig. 3-a   d = 1

Energy Density, Mutter Field and Magnetic Field of two vortices for $\lambda = 1$ at various seperations.

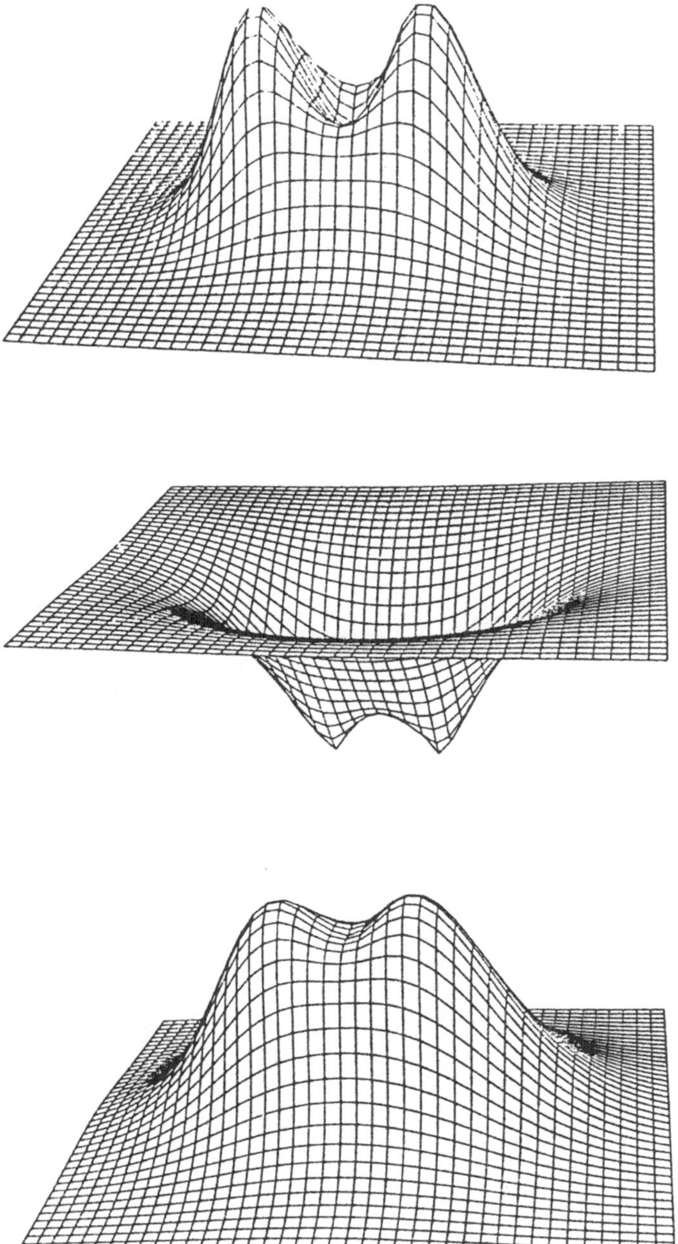

Fig. 3-b

d = 2

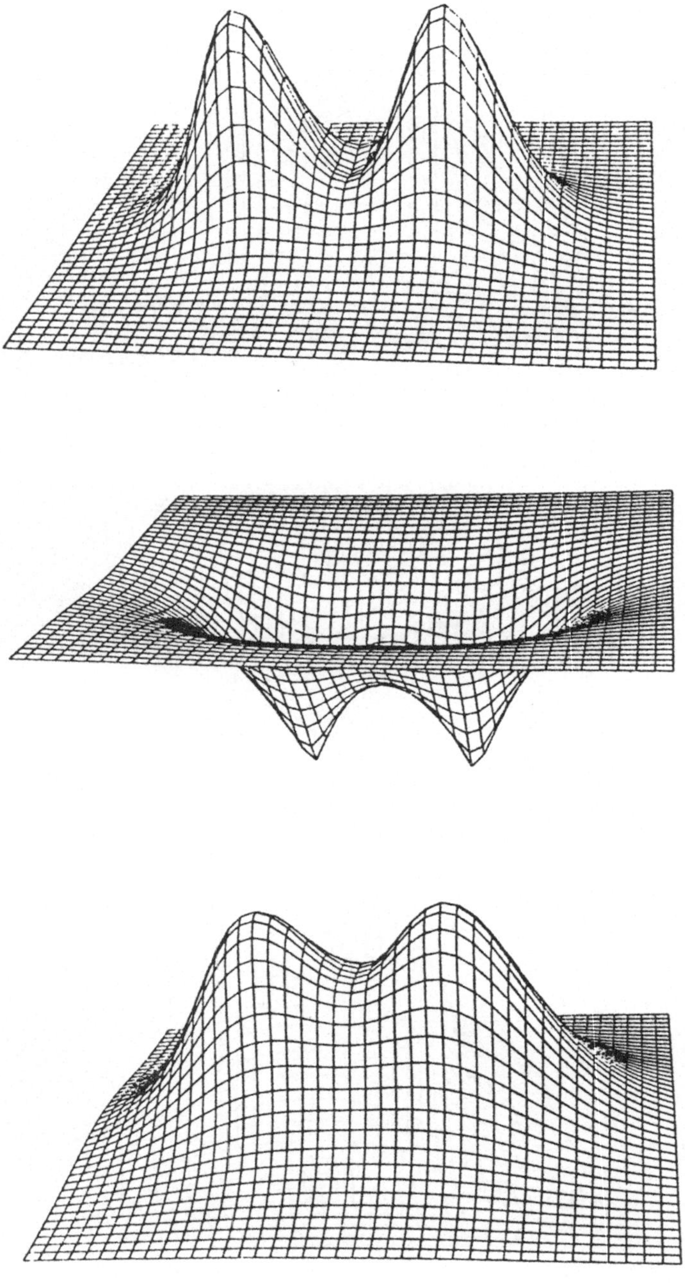

Fig. 3-c

$d = 3$

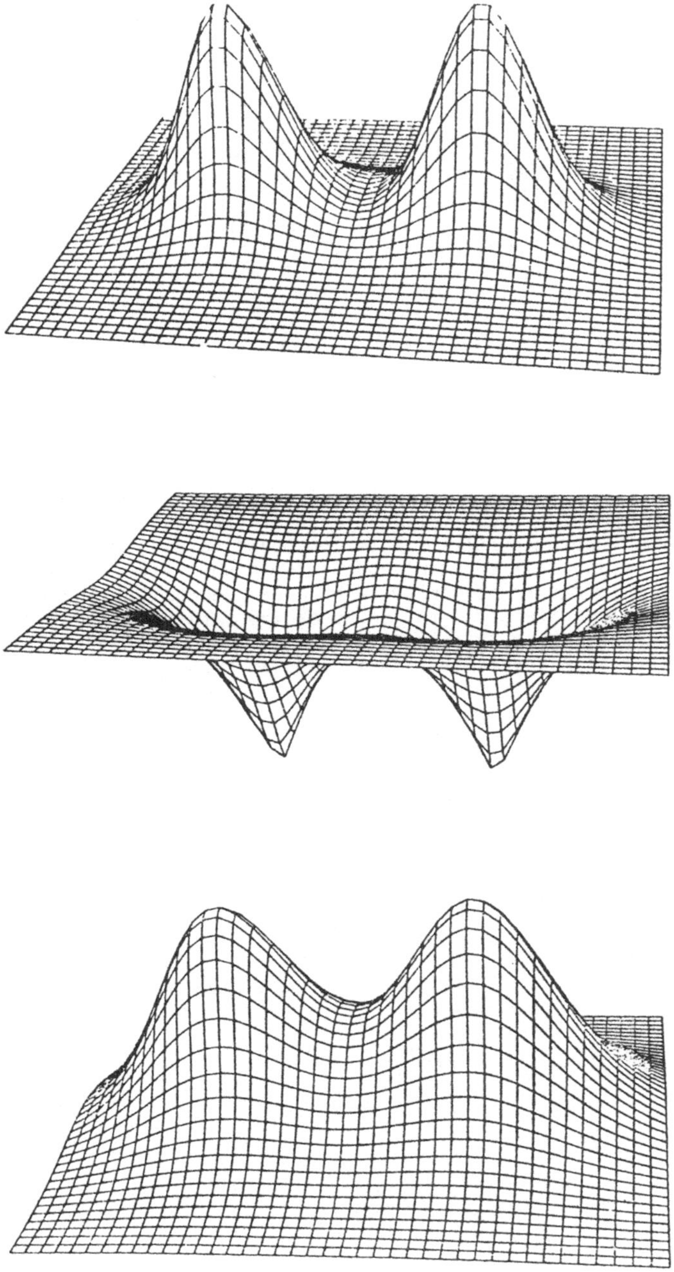

Fig. 3-d

d = 4

d	λ =0.7	λ =1.0	λ =1.3
0	1.653	2.000	2.309
1	1.653	2.000	2.308
2	1.656	2.000	2.299
3	1.665	2.000	2.276
4	1.680	2.000	2.254
5	1.696	2.000	2.242
6	1.710	2.000	2.236
7	1.718	2.000	2.234
8	1.723	2.000	2.233
∞	1.728	2.000	2.232

The following Figures (3a $\div$ d) illustrate the profiles of the energy density, matter field and magnetic field for λ =1 and various separations of the vortices.

Clearly, the interaction between vortices is always attractive for λ=0.7 and repulsive for λ=1.3. The value λ=1 also emerges as a critical value, for which the vortices appear to be in equilibrium at any separation. The point λ=1 is indeed of particular physical and mathematical significance. For this critical value of the coupling constant it is possible to derive a lower bound on the energy, additive in the topological quantum number q [7]. The bound is saturated when a set of first order non-linear partial differential equations in the fields A_i and ϕ is satisfied; then, of course, also the second order Euler-Lagrange equations are fulfilled. The numerical analysis indicates that the bound can be saturated with two vortices at arbitrary separations. More recent analytical developments have shown that, always for λ=1, solutions with any number of vortices at arbitrary locations in the plane do indeed exist. These developments will be the object of my second lecture. To conclude the present one, let me mention that in superconductors the value λ =1 separates material exhibiting superconductivity of the first ($\lambda < 1$) or second ($\lambda > 1$) type. In type I superconductors a magnetic field is expelled to the surface, whereas in type II superconductors it penetrates the material, forming an array of tubes of flux. This behavior is easily understood keeping in mind the tendency of vortices to coalesce ($\lambda < 1$) or separate ($\lambda > 1$) and considering the geometry of the system.

III. Multi-vortex Configurations for λ = 1

When λ=1 it is possible to derive a lower bound on the energy functional [7]. The first term in the r.h.s. of Eq. (2.6) is integrated by parts to express it as the sum of the second term in the same equation plus a remainder, involving the strength of the magnetic field and the density of the matter field:

$$\int dz d\bar{z}\, [(\partial - iA)\phi]\, [(\bar{\partial} - i\bar{A})\bar{\phi}] = \int dz d\bar{z}\, [(\bar{\partial} - i\bar{A})\phi]\, [(\partial - iA)\bar{\phi}] -$$

$$- i \int dz d\bar{z} (\partial\bar{A} - \bar{\partial}A)\phi\bar{\phi}. \qquad (3.1)$$

Using Eq. (3.1) the energy functional is rewritten as

$$\mathcal{E} = \frac{1}{\pi} \, dz d\bar{z}\, [\, |(\partial - i\bar{A})\phi|^2 +$$

$$+ |\bar{\partial}A - \partial\bar{A}|^2 + \frac{1}{16} (|\phi|^2 - 1)^2 - \frac{i}{2} (\partial\bar{A} - \bar{\partial}A)\phi\bar{\phi}\,]. \qquad (3.2)$$

Adding and subtracting the integral

$$-\frac{i}{2\pi} \int dz d\bar{z}\, (\partial\bar{A} - \bar{\partial}A) = q, \qquad (3.3)$$

giving the topological quantum number, the last three terms in the r.h.s. of Eq. (3.2) are combined into a perfect square (it is crucial for this that $\lambda = 1$), arriving finally at the equation

$$\mathcal{E} = \frac{1}{\pi}\int dz d\bar{z}\, \{ |(\partial - i\bar{A})\phi|^2 +$$

$$[\tfrac{1}{4}(|\phi|^2 - 1) - i(\partial\bar{A} - \bar{\partial}A)]^2\} + q. \tag{3.4}$$

It is apparent now that $\mathcal{E}$ is bounded below by q, the bound being saturated if and only if the equations

$$(\bar{\partial} - i\bar{A})\phi = 0 \tag{3.5}$$

and

$$\partial\bar{A} - \bar{\partial}A + \frac{i}{4}(|\phi|^2 - 1) = 0 \tag{3.6}$$

are satisfied. (If q is negative, an analogous procedure allows to establish the bound $\mathcal{E} \geq -q$.)

In Reference [4], it was shown that Eqs. (3.5) and (3.6) may be solved in a rotationally symmetric configuration of q superimposed vortices. The numerical results obtained in Reference [5] led to the conjecture that solutions with q vortices at arbitrary positions in the plane do exist. In a most recent development this conjecture was demonstrated true [9]. In between, a study of the deformation problem made by E. Weinberg [8] showed that the solutions of Eqs. (3.5), (3.6), if they exist, belong to a 2q parameter class (as one would indeed expect, if the whole arbitrariness is in the location of the vortices). This nice analysis proceeds via the derivation of an Atiyah - Singer index theorem and it is worthwhile considering it in some detail.

Eqs. (3.5) and (3.6) admit as trivial infinitesimal deformation any gauge transformation of the fields. To eliminate these a Coulomb gauge condition is imposed. The complete set of equations (including also the complex conjugate of Eq. (3.5) becomes then

$$(\bar{\partial} - i\bar{A})\phi = 0, \tag{3.7a}$$

$$(\partial + iA)\bar{\phi} = 0, \tag{3.7b}$$

$$\partial\bar{A} - \bar{\partial}A + \frac{i}{4}(\phi\bar{\phi} - 1) = 0, \tag{3.7c}$$

$$\partial\bar{A} + \bar{\partial}A = 0. \tag{3.7d}$$

We assume now that ϕ, $\bar{\phi}$, A and $\bar{A}$ represent the fields of an actual solution to Eqs. (3.7a) - (3.7d) and inquire what infinitesimal changes $\delta\phi$, $\delta\bar{\phi}$, δA, $\delta\bar{A}$ (if any) would maintain the equations satisfied. Putting together the four variations into a single vector ξ, it is immediate to show that the infinitesimal deformations should satisfy the equation

$$\mathcal{D}\xi = 0, \tag{3.8}$$

where $\mathcal{D}$ is the matrix-differential operator

$$\mathcal{D} = \begin{pmatrix} \bar\partial - i\bar A & 0 & 0 & -i\phi \\ 0 & \partial + iA & i\bar\phi & 0 \\ \frac{i\phi}{4} & \frac{i\phi}{4} & -\bar\partial & \partial \\ 0 & 0 & \bar\partial & \partial \end{pmatrix} \tag{3.9}$$

The index I of $\mathcal{D}$ is defined as the number of solutions of Eq. (3.8) less the number of solutions of the equation

$$\mathcal{D}^+\xi = 0, \tag{3.10}$$

where $\mathcal{D}^+$ is the adjoint of $\mathcal{D}$. Since it is rather easy to prove (from positivity) that Eq. (3.10) has no solutions, the index of $\mathcal{D}$ gives also the number of infinitesimal deformations.

All solutions of Eq. (3.8) or (3.9) are also solutions of the equation

$$\mathcal{D}^+\mathcal{D}\xi = 0 \tag{3.11}$$

or, respectively,

$$\mathcal{D}^+\mathcal{D}\xi = 0. \tag{3.12}$$

On the other hand the hermitian operators $\mathcal{D}^+\mathcal{D}$ and $\mathcal{D}\mathcal{D}^+$ have the same non-vanishing eigenvalues, because $\mathcal{D}^+\mathcal{D}\xi = c\xi$ with $c \neq 0$ implies $\mathcal{D}\mathcal{D}^+\mathcal{D}\xi = c\,\mathcal{D}\xi$ with non-vanishing $\mathcal{D}\xi$. The index may therefore be expressed as a trace,

$$I = \mathrm{Tr}\left(\frac{M^2}{\mathcal{D}^+\mathcal{D} + M^2} - \frac{M^2}{\mathcal{D}\mathcal{D}^+ + M^2} \right) , \tag{3.13}$$

taken both in function and matrix space.

M^2 in Eq. (3.13) is a free parameter, which can be assigned any value. It is then convenient to evaluate I from Eq. (3.13) by letting $M \to \infty$.

Simple algebra gives

$$\begin{aligned} \mathcal{D}^+\mathcal{D} &= -\partial\bar\partial + \mathcal{L} \\ \mathcal{D}\mathcal{D}^+ &= -\partial\bar\partial + \overline{\mathcal{L}} \end{aligned} \tag{3.14}$$

where $\mathcal{L}$ and $\overline{\mathcal{L}}$ are operators linear in the derivatives. Expanding

$$\frac{M^2}{\mathscr{D}^\dagger\mathscr{D}+M^2} = \frac{M^2}{-\partial\bar\partial + M^2} - \frac{M^2}{-\partial\bar\partial + M^2}\mathscr{L}\frac{M^2}{-\partial\bar\partial + M^2} + \ldots \tag{3.15}$$

and inserting into the trace, it is possible to verify by diagrammatic techniques that terms of higher order do not contribute to I in the limit $M \to \infty$. It then follows

$$I = \lim_{M\to\infty} M^2 \int d^2x \, \langle x|(-\partial\bar\partial + M^2)^{-1} \, \mathrm{tr}(-\mathscr{L}+\tilde{\mathscr{L}})(-\partial\bar\partial + M^2)^{-1}|x\rangle, \tag{3.16}$$

where the trace has been resolved into a sum over position eigenfunctions and a trace over vector indices only. But

$$\mathrm{tr}(-\mathscr{L}+\tilde{\mathscr{L}}) = -2i(\partial\bar{A} - \bar\partial A) \tag{3.17}$$

and introducing the Fourier representation of $(-\partial\bar\partial + M^2)^{-1}$ one finally arrives at

$$I = \lim_{M\to\infty} \int dz d\bar{z}\{ -2i(\partial\bar{A} - \bar\partial A)\} \times \int \frac{d^2k}{2\pi^2} \frac{M^2}{(k^2 + M^2)^2} = 2q. \tag{3.18}$$

To demonstrate that Eqs. (3.7a) ÷ (3.7d) have arbitrary multivortex solutions, it is convenient to solve the gauge condition (3.7d) expressing A and $\bar{A}$ in terms of a real function:

$$A = i\partial\psi, \ \bar{A} = -i\bar\partial\psi. \tag{3.19}$$

Eqs. (3.7a) and (3.7b) reduce then to the statement that

$$f \equiv e^{-\psi}\phi$$

is an analytic function of z. Indeed, substituting Eq. (3.19) into Eq. (3.7a) we find

$$(\bar\partial - \bar\partial\psi)\phi = e^{\psi}\bar\partial(e^{-\psi}\phi) = 0,$$

i.e. $\frac{\partial f}{\partial \bar{z}} = 0, \ f = f(z).$

Setting

$$\phi = e^{\psi} f \tag{3.20}$$

the remaining equation becomes

$$\partial\bar\partial\psi = \frac{1}{8}(e^{2\psi} f\bar{f} - 1) \tag{3.21}$$

and, with the further substitution

$$\psi = \chi - \tfrac{1}{2}\ln f - \tfrac{1}{2}\ln \bar{f}, \tag{3.22}$$

we finally arrive at the equation

$$\partial\bar{\partial}\chi = \frac{1}{8}(e^{2\chi} - 1). \qquad (3.23)$$

The boundary conditions for χ follow from Eqs. (3.20), (3.22) and the behavior of ϕ. χ must approach 0 at infinity and diverge as $\ln |z - z_i|$ at the vortices where f(z) has a zero. The singular behavior of χ at the vortices is the clue to the existence of multi-vortex solutions. Indeed, because of the allowed divergences at $z = z_i$, Eq. (3.23) should more properly be written as

$$\partial\partial\chi = \frac{1}{8}(e^{2\chi} - 1) + \pi\sum_i\delta^{(2)}(z - z_i, \bar{z} - \bar{z}_i). \qquad (3.24)$$

Given the presence of sources in the r.h.s. of Eq. (3.24) the existence of solutions for arbitrary locations of the vortices is not surprising. (Think of the Poisson equation $\partial\bar{\partial}\chi = \pi\sum_i\delta^{(2)}(z - z_i, \bar{z} - \bar{z}_i)$). Of course, the equation is not linear and so to prove the existence of solutions remains non-trivial. To obtain such a proof it is convenient to express χ as sum of a zero order Ansatz with the correct singularities and a finite correction $\delta\chi$. The equation satisfied by $\delta\chi$ is the Euler-Lagrange equation for the minimization of a reduced energy functional $\delta\mathcal{E}$. Given the good convexity properties of $\delta\mathcal{E}$ and the regular behavior of $\delta\chi$, theorems of functional analysis can be used to establish the existence of a unique minimum. The proof has been recently given by Taubes [9].

In summary, for $\lambda = 1$ the properties of a system of interacting vortices have been completely elucidated. The existence of a lower bound on the energy and of a change of variables by which finite physical fields are expressed in terms of a superpotential with allowed singularities plays a fundamental role in the mathematical analysis of the system. It is interesting to notice that also in other field theories, in spaces of higher dimensionality, the same ingredients allow to prove the existence of non-interacting, localized structures. For $\lambda \neq 1$ the system is not amenable to an analytic solution. However, through numerical and asymptotic results a good understanding of the properties of interacting vortices has been achieved.

References

1. V. L. Ginzburg and L. D. Landau, Zh. Eksp. Teor. Fiz. 20, 1064 (1950).

2. H. B. Nielsen and P. Olesen, Nucl. Phys. B 61, 45 (1973).

3. A. A. Abrikosov, Zh. Eksp. Teor. Fiz. 32, 1442 (1957) [Sov. Phys. JETP 5, 1174 (1957)]; L. P. Gor'kov, Zh. Eksp. Teor. Fiz. 34, 734 (1958); ibid, 36, 1918 (1959); [Sov. Phys. JETP 7, 505 (1958); ibid, 9, 1364 (1959)].

4. H. J. de Vega and F. A. Schaposnik, Phys. Rev. D 14, 1100 (1976).

5. L. Jacobs and C. Rebbi, Phys. Rev. B 19, 4486 (1979).

6. E. Müller-Hartmann, Phys. Lett. 23, 521 (1966); ibid, p. 619; E. B. Bogomol'nyi, Sov. J. Nucl. Phys. 23, 588 (1976).

7. L. Kramer, Phys. Rev. B 3, 3821 (1971); E. B. Bogomol'nyi, Sov. J. Nucl. Phys. 24, 449 (1976).

8. E. Weinberg, Columbia University preprint (1979).

9. C. Taubes, Harvard University preprint (1979).

ON GROUPS OF GAUGE TRANSFORMATIONS

Andrzej Trautman
Institute of Theoretical Physics
Warsaw University
Hoza 69, Warszawa, Poland

0. Summary

Groups of gauge transformations (gauge groups) are defined in the framework of principal bundles. The gauge group of a trivial bundle is exhibited and the gauge aspect of gravitation is compared to that of Yang-Mills theories.

1. Introduction

Recent developments in theoretical physics indicate a wide-ranging importance of gauge fields. There are reasons to believe that all fundamental forces are mediated by particles which are quanta of appropriate gauge fields. In the approximation of classical physics, gauge configurations are described best by connections on principal bundles over spacetime. The mathematical framework of fibre bundles provides precise definitions of the notions used in classical gauge theories. Among them are the notions of gauge transformations. In the older literature, a distinction was made between transformations of the first and second kind [1], whereas in recent works one refers to global and local gauge transformations [2]. There is also considerable interest in gauge transformations in the theory of gravitation [3] and its 'supersymmetric' modification [4].

Extending an earlier note [5], this paper contains the definitions and elementary properties of gauge groups. The theory of gravitation is contrasted to a Yang-Mills theory over Minkowski spacetime. The paper follows the standard notation and terminology used in differential geometry and applications of fibre bundle theory to physics (see, for example, [6] and the references given therein). All manifolds and maps are assumed to be of class C^∞. A principal fibre bundle includes in its definition a projection π of the total space of the bundle P on the base M and an action of a Lie group G on P to the right. The action is free and transitive on the fibres of π. If $\delta:P\times G \to P$ is the map defining the action, then one writes $\delta(p,a) = \delta_a(p)$ or pa, for simplicity. A connection is given by a one-form ω on P, with values in the Lie algebra of G. A (local) section s of π, $s:M \to P$, $\pi\circ s = id_M$, corresponds to the physicists' idea of choosing a gauge. Let $\rho:G\times N \to N$ be a map defining a (left) action of G in a manifold N. A (generalized) Higgs field of type ρ is a map $\varphi:P \to N$, equivariant under the action of G, i.e. such that $\varphi\circ\delta_a = \rho_{a^{-1}}\circ\varphi$, where $\rho_a(n) = \rho(a,n)$, $a \in G$ and $n \in N$. The pullback $A = s^*\omega$ is the gauge potential in that gauge.

2. Automorphisms of principal bundles

A diffeomorphism $u:P \to P$ is an automorphism of the bundle $\pi:P \to M$ if there is a diffeomorphism $v:M \to M$ such that $\pi \circ u = v \circ \pi$ and $u(pa) = u(p)a$ for any $p \in P$ and $a \in G$. The set AutP of all automorphisms of P is a group under composition of maps. The diffeomorphism v is uniquely determined by the automorphism u and there is a homomorphism of groups $j:\mathrm{AutP} \to \mathrm{DiffM}$ given by $j(u) = v$. An automorphism u is called vertical (or based) if $j(u) = id_M$; the set Aut_0P of all vertical automorphisms is a normal subgroup of AutP and the sequence

$$1 \to \mathrm{Aut}_0 P \to \mathrm{AutP} \xrightarrow{j} \mathrm{DiffM}$$

is exact. There are natural bijections among the following three sets

(i) Aut_0P;

(ii) the set of all maps $U:P \to G$ such that $U(pa) = a^{-1}U(p)a$ for any $p \in P$ and $a \in G$;

(iii) the set of all sections of the bundle $E \to M$, associated to $\pi:P \to M$ by the adjoint action of G on itself [7].

The correspondence between (i) and (ii) is given by

$$u(p) = pU(p).$$

Let $k:P\times G \to E$ be the canonical map, $k(p,a) = k(pb,b^{-1}ab)$, where $a,b \in G$. If U is as in (ii), then $k(p,U(p))$ depends only on $\pi(p)$ and defines a section $\tilde{u}$ of $E \to M$.

Example 1.

If c is a central element of G, then the constant map $U:P \to G$, $U(p) = c$, defines a vertical automorphism.

Example 2.

Let M be n-dimensional, $x \in M$ and let T_xM be the tangent space to M at x. A (linear) frame at x is a (linear) isomorphism $e:R^n \to T_xM$. The set LM of all such frames at all points of M gives rise, in a natural manner, to the principal bundle of frames; $\pi(e) = x$. The action of its group, GL(n,R), is by composition of linear maps, $ea = e \circ a$. The bundle E associated to LM by ad consists of all linear automorphisms of the tangent spaces to M; a section of $E \to M$ is a field of invertible tensors of mixed valence on M. Any vertical automorphism of LM is given by a tensor field of this kind.

Example 3.

Let $T_xv:T_xM \to T_{v(x)}M$ denote the tangent map to $v:M \to M$ at $x \in M$. For any $v \in \mathrm{DiffM}$, one defines its lift $Lv:LM \to LM$ by

$$Lv(e) = T_xv \circ e, \quad \text{where } x = \pi(e). \tag{1}$$

Clearly, Lv is an automorphism of $LM \to M$ and $jL = id_{\mathrm{DiffM}}$. For any $u \in \mathrm{AutLM}$, the composition $u \circ (Lj(u^{-1}))$ is a vertical automorphism.

For any manifold M and (Lie) group G, one defines the group G^M of all maps from M to G; the composition in G^M is induced pointwise from G. There is a natural homomorphism τ of DiffM into the automorphism group of G^M given by

$$\tau_v(w) = w \circ v^{-1}, \quad \text{where } v \in \text{DiffM}, \quad w:M \to G.$$

Proposition 1.

The group of all automorphisms of the trivial bundle $pr_1: M\times G \to M$ is isomorphic to the semi-direct product of DiffM and G^M relative to τ.

Indeed, any automorphism $u: M\times G \to M\times G$ may be represented by the pair (v,w), where $v = j(u) \in \text{DiffM}$ and $w: M \to G$ is such that

$$u(x,a) = (v(x), w(v(x))a) \quad \text{for any } x \in M \text{ and } a \in G.$$

Moreover, if the automorphism u' is represented in this way by (v',w'), then

$$u \circ u' \text{ is represented by } (v \circ v', \; w \circ \tau_v(w')).$$

3. Gauge groups and symmetrics

In any physical theory, besides dynamical variables which are subject to equations of motion, there occur absolute elements, such as external forces or the metric tensor in special relativity. In a gauge theory, the absolute elements are often given by geometric objects, defined on the bundle $\pi: P \to M$, in addition to the connection and the Higgs field which play a dynamical role. It is reasonable to define the gauge group of such a theory as the subgroup $\mathcal{G}$ of AutP, consisting of all automorphisms of π which preserve the absolute elements. The elements of $\mathcal{G}$ are called gauge transformations. A pure gauge transformation is a vertical element of $\mathcal{G}$. The pure gauge group

$$\mathcal{G}_0 = \mathcal{G} \cap \text{Aut}_0 P$$

is a normal subgroup of $\mathcal{G}$ and there is the exact sequence

$$1 \to \mathcal{G}_0 \to \mathcal{G} \to \mathcal{G}/\mathcal{G}_0 \to 1. \qquad (2)$$

Gauge transformations act on sections and connections: if s is a section of $\pi: P \to M$ and $u \in \mathcal{G}$, then $s' = u \circ s \circ v^{-1}$ is another section. Similarly, the pullback $\omega' = u^*\omega$ of a connection form is another connection form and there is the equality of potentials

$$s^*\omega' = v^* s'^* \omega.$$

This can be interpreted as follows: the form ω describes the same geometry and physics as ω' does, only 'translated' by the diffeomorphism v. In other words, any invariant constructed from ω' and the absolute elements at $x \in M$ is equal to the corresponding invariant constructed from ω at $v(x)$.

The gauge group of a Yang-Mills theory over Minkowski space is easily obtained on the basis of Proposition 1: the group $\mathcal{G}_0$ is isomorphic to G^M, whereas $\mathcal{G}$ is isomorphic to the semi-direct product of the inhomogeneous Lorentz group and G^M relative to τ.

The following example shows that the gauge exact sequence need not split for a non-trivial bundle.

Example 4.
Consider the Z-bundle $\pi: R \to U(1)$, $\pi(t) = \exp 2\pi i t$, and assume its total space R to have the standard metric and orientation. If these two elements are considered as absolute, then G reduces to R, the group of translations, and the sequence (2) becomes

$$1 \to Z \to R \to U(1) \to 1.$$

By definition, a diffeomorphism $v: M \to M$ is a symmetry of a gauge configuration given by ω on P if there is a gauge transformation $u: P \to P$ which covers v, i.e. $j(u) = v$, and

$$u^*\omega = \omega,$$

Similarly, a Higgs field of type ρ given by the map $\varphi: P \to N$ admits v as a symmetry if it is invariant under $u \in G$,

$$u^*\varphi = \varphi,$$

and $j(u) = v$. If N is an orbit of G, then φ restricts the bundle $\pi: P \to M$ to the little group of $\varphi_0 \in N$,

$$H = \{a \in G: \rho(a, \varphi_0) = \varphi_0\}.$$

The total space Q of the restricted bundle is

$$Q = \{p \in P: \varphi(p) = \varphi_0\}$$

and it is straightforward to prove

Proposition 2.
A Higgs field with values in an orbit of G is invariant under $u \in \text{Aut}P$ if and only if $u \in \text{Aut}Q$.

4. Gravitation

The 'kinematic' aspect of gravitation is described by a connection ω on the bundle LM of linear frames of an n(=4)-dimensional manifold and by a metric which may be considered as a generalized Higgs field $g: LM \to N \subset L^2_s(R^n, R)$, where N is an orbit of GL(n,R) in the space of symmetric, n n matrices. According to the theorem on inertia of quadratic forms on R^n there is a one-to-one correspondance between the set of all such orbits and the collection of all possible signatures of these forms. The 'dynamics' consists of differential equations for ω and g.

An important aspect of gravitation is the 'concrete' nature of LM: its elements are linear frames on M whereas not much can be said about the elements of an 'abstract' bundle P. The bundle $\pi: LM \to M$ is 'richer' than an abstract bundle. Its additional structure is completely described by the canonical one-form $\theta: TLM \to R^n$ defined by

$$\theta_e = e^{-1} \circ T_e \pi \qquad (3)$$

where θ_e is the restriction of θ to $T_e LM$ and $e \in LM$ is interpreted as an isomorphism from R^n to $T_{\pi(e)}M$. Clearly, $\theta_{ea} \circ T_e \delta_a = a^{-1} \circ \theta_e$, thus proving

$$\delta_a^* \theta = a^{-1} \circ \theta \qquad (4)$$

and, for any $u \in TLM$,

$$\theta(u) = 0 \Leftrightarrow T\pi(u) = 0. \qquad (5)$$

Proposition 3.

A principal bundle $\pi : P \to M$, with an n-dimensional base and structure group $GL(n,R)$ is isomorphic to the bundle of linear frames $LM \to M$ if and only if there is a map $\theta : TP \to R^n$, linear on the fibres of $TP \to P$, and satisfying (4) and (5) for any $a \in GL(n,R)$ and $u \in TP$.

Indeed, if there is such a θ on P, then the (based) isomorphism $h : P \to LM$ is determined as follows. Condition (5) means that, for any $p \in P$, the linear map $T_p\pi : T_pP \to T_{\pi(p)}M$ factors through $\theta_p : T_pP \to R^n$, i.e. there is a linear map

$$h(p) : R^n \to T_{\pi(p)}M$$

such that $h(p) \circ \theta_p = T_p\pi$. The map $h(p)$ is uniquely defined; moreover, it is an isomorphism and, therefore, an element of LM lying over $\pi(p)$. The equivariance of h follows from (4).

The canonical ('soldering') form θ plays the role of an absolute element in the theory of gravitation. The following two propositions are useful in determining the gauge groups in gravity:

Proposition 4.

If $u \in Aut_0 LM$ and $u^*\theta = \theta$, then $u = id$. This follows directly from the definition (3) of θ: $u^*\theta = \theta$ is equivalent to $\theta_{u(e)} \circ T_e u = \theta_e$, any e. Using (3) and $\pi \circ u = \pi$, one obtains $u(e) = e$.

Proposition 5.

If $u : LM \to LM$ is a diffeomorphism such that $\pi \circ u = v \circ \pi$ for some diffeomorphism $v : M \to M$, then the following conditions are equivalent:

(i) $u = Lv$,

(ii) $u^*\theta = \theta$.

Indeed, it follows from the definition of θ and (1) that

$$(u^*\theta)_e = u(e)^{-1} \circ Lv(e) \circ \theta_e \quad \text{for any } e \in LM.$$

(i) $\Rightarrow$ (ii) is now obvious and (ii) $\Rightarrow$ (i) follows from the surjectivity of θ_e.

Let Ω and Θ be, respectively, the curvature and torsion two-forms of a linear connection ω. Denoting by Ω' and Θ' the forms corresponding to $\omega' = u^*\omega$, $u \in AutLM$, one obtains

Proposition 6.

For any $u \in \mathrm{AutLM}$,

$$\Omega' = u^*\Omega.$$

If, moreover, $u = Lv$, then

$$\Theta' = u^*\Theta. \tag{6}$$

It is important to realize that (6) does not, in general, hold unless u is the lift of a diffeomorphism; one can 'generate torsion' by applying a suitable vertical automorphism to a symmetric connection. These remarks are intended to justify our definition of the group G of gauge transformations in theories of gravity based on LM:

$$G = \{u \in \mathrm{AutLM}: u^*\theta = \theta\}.$$

By Proposition 5 this group is isomorphic to DiffM and by Proposition 4 the group G_0 of pure gauge transformations reduces to {id}. This should be contrasted with the case of a Yang-Mills theory over Minkowski space, for which $G_0 = G^M$ is 'large' and G/G_0 is finite-dimensional ('small').

Incidentally, the lift $L:\mathrm{DiffM} \to \mathrm{AutLM}$ defines a splitting of the sequence

$$1 \to \mathrm{Aut}_0\mathrm{LM} \to \mathrm{AutLM} \to \mathrm{DiffM} \to 1$$

and the representation $u \to (v, u \circ Lv^{-1})$, $v = j(u)$, yields an isomorphism of AutLM on the semi-direct product of DiffM and $\mathrm{Aut}_0\mathrm{LM}$, relative to the homomorphism $\sigma:\mathrm{DiffM} \to \mathrm{Aut}(\mathrm{Aut}_0\mathrm{LM})$, where $\sigma_v(w) = Lv \circ w \circ Lv^{-1}$ for $w \in \mathrm{Aut}_0\mathrm{LM}$.

Acknowledgement

This note reproduces a part of my lecture given at the Symposium on Topological and Geometrical Methods in Gauge Theory, held in September 1979 at McGill University in Montreal. I gratefully acknowledge the hospitality and financial support extended to me by the organizers of the Symposium. Conversations with John Harnad, held in both Warsaw and Montreal, influenced the section on Gauge Groups and Symmetries.

References

[1] N.N.Bogoliubov and D.V.Shirkov, Introduction to the theory of quantized fields, Interscience, New York 1959.

[2] J.C.Taylor, Gauge theories of weak interactions, Cambridge University Press, Cambridge 1976.

[3] R.Utiyama, Phys.Rev. 101 (1956) 1597;
T.W.B.Kibble, J.Math.Phys. 2 (1961) 212;
E.A.Lord, Proc.Camb.Philos.Soc. 69 (1971) 423;
C.N.Yang, Phys.Rev.Lett. 33 (1974) 445;
F.W.Hehl et al., Rev.Mod.Phys. 48 (1976) 393;
Y.M.Cho, Phys.Rev. D14 (1976) 2521;

A.Trautman, article in the GRG Einstein Volume, ed.A.Held et al., Plenum Press, New York 1980.

[4] D.Z.Freeman, P.van Nieuwenhuizen and S.Ferrara, Phys.Rev. D13 (1976) 3214.

[5] A.Trautman, Bull.Acad.Polon.Sci., sér.sci.phys.et astron.27 (1979) 7.

[6] A.Trautman, Czech.J.Phys.B29 (1979) 107.

[7] J.Dieudonné, Eléments d'analyse, t.3 (§16.14), Gauthier-Villars, Paris 1970.

CONSTRUCTION OF GAUGE FIELDS FROM INITIAL DATA

James A. Isenberg
Department of Applied Mathematics
University of Waterloo
Waterloo, Ontario
Canada
N2L 3G1

Whether or not the non-Abelian gauge theories are found to be useful in describing the physical world, it is unlikely that they have any measurable effects at the level of classical physics. Still, it is useful to study these theories in a purely classical way since the classical analysis is the first step to quantum understanding as well as to quantum calculations[1].

The initial value (or "3+1") method is a particularly useful way to classically analyze a given field theory for two reasons. Firstly, it is the most direct way to check whether or not a particular theory is "consistent" (in the sense that proper solutions exist) and to count its degrees of freedom. Secondly, it provides a straightforward procedure for constructing and comparing lots of spacetime solutions for most (consistent) field theories. We might add that the 3+1 analysis is a necessary first step if one wants to quantize any theory canonically.

What is the 3+1 formulation of a theory? The idea is that rather than discussing a classical field theory via

> <u>spacetime-covariant formulation</u>: spacetime-covariant fields and spacetime field equations on a bundle over a 4-dimensional Lorentzian manifold

one uses

> <u>3+1 formulation</u>: space-covariant fields and constraint equations on a bundle over a 3-dimensional Riemannian manifold with evolution equations involving gauge evolvers.

So for a given classical field theory stated in spacetime convariant form, there are two parts to its 3+1 analysis: (a) The transformation from spacetime form to 3+1 form (in the process of which one checks consistency and then counts degrees of freedom); (b) The construction of spacetime solutions via evolution from chosen initial data.

Both parts of the initial value analysis are surveyed-in rather general terms - in some recent review articles[2]. Here, we discuss its application to gauge fields. We concentrate on the Einstein-Yang-Mills (EYM) theory which admits a very nice (and useful) 3+1 formulation (see §I). Using EYM as an example, we point out (see §II) how the 3+1 spacetime builder specifies the gauges and bundles of the spacetime solution he is building. While the 3+1 formulation of gauge theories other than EYM is often more difficult both to obtain and to use, there has been some recent success. We discuss two examples: (1) The dynamic torsion theory of FAIRCHILD [1977] and YASSKIN [1978] is shown by 3+1 analysis to be consistent for generic initial data and to have 8 degrees of freedom (see §III); (2) The spacelike characteristic surfaces which occur in HORNDESKI'S [1976] generalized electromagnetic theory are shown by 3+1 solution construction to be less malignant than feared, at least in simple cases (see §IV).

I. Obtaining the 3+1 Formulation of the Einstein-Yang-Mills Theory

In spacetime covariant form[3], the EYM fields are described by a pseudo-Riemannian metric $\mathbb{g}_{\mu\nu}$ on a spacetime manifold M, and by a connection $\mathbb{A}^{D}_{\ \mu}$ on a G-bundle B_M over M (any semisimple group G). An EYM solution is specified by a choice of B_M, M, $\mathbb{g}_{\mu\nu}$ and $\mathbb{A}^{D}_{\ \mu}$ such that

$$\mathbb{D}_\mu \mathbb{F}^{D\mu\nu} = 0 \tag{1}$$

and

$$\mathbb{G}_{\mu\nu} = \frac{\lambda}{2}\left(\mathbb{F}^{A}_{\ \mu}{}^{\alpha}\, \mathbb{F}_{A\alpha\nu} - \frac{1}{4}\mathbb{g}_{\mu\nu}\, \mathbb{F}_{A}^{\ \alpha\beta}\, \mathbb{F}^{A}_{\ \alpha\beta}\right) \tag{2}$$

are satisfied everywhere on M. Here $\mathbb{D}_\mu$ is the spacetime and G-covariant derivative involving both $\mathbb{A}^{D}_{\ \mu}$ and the Riemannian connection $\mathbb{\Gamma}^{\alpha}{}_{\beta\mu}$ (metric-compatible, torsion free); $\mathbb{F}^{D}_{\ \mu\nu}$ is the G-curvature (or "Yang-Mills field") built from $\mathbb{A}^{D}_{\ \mu}$, and $\mathbb{G}_{\mu\nu}$ is the Einstein tensor built from $\mathbb{\Gamma}^{\alpha}_{\ \beta\mu}$ and $\mathbb{g}_{\mu\nu}$. Also, λ is the coupling constant.

To obtain the 3+1 formulation of the EYM theory, we follow the step-by-step Bergmann-Dirac procedure[4].

1) *Field Decomposition*: Let S be a spacelike surface in M and choose coordinates (t,x^a) to be compatible with S in the sense that S is a t = const. surface. One may then write out the metric in the form

$$\mathbb{g} = \gamma_{ab}(dx^a + M^a dt)(dx^b + M^b dt) - N^2 dt^2 \tag{3}$$

so that $\mathbb{g}$ is completely parametrized by the S-covariant fields γ_{ab} ("intrinsic metric"), M^a ("shift" vector), and N ("lapse" scalar). To decompose the Yang-Mills connection (and any other fields which may be around), it is convenient to work with the surface-compatible basis[5] $\{e_\perp, \frac{\partial}{\partial x^a}\}$ and its dual $\{\theta^\perp, \theta^a\}$, where $e_\perp$ is the unit (future-pointing) vector normal to S. One may then write

$$\mathbb{A}^B = A^B_{\ \perp}\theta^\perp + A^B_{\ m}\theta^m \ . \tag{4}$$

Just as γ_{ab} specifies an intrinsic (Riemannian) metric for S, so $A^B_{\ a}$ specifies an intrinsic G-connection in a bundle B_S over S. These define a unique spatial G-covariant derivative (metric compatible, torsion free) D_m. Using this D_m, together with the Lie derivative $\mathcal{L}_{\mathbb{Y}}$ in directions $\mathbb{Y}$ off the surface[6], one may decompose the spacetime G-covariant derivative $\mathbb{D}_m$ as follows: [Here $\mathbb{V}^\mu \leftrightarrow (V^m, V^\perp)$ is any G-representation-valued spacetime vector (and its 3+1 decomposition) with the G-indices suppressed]

$$\mathbb{D}_{e_\perp}\mathbb{V}^\perp = \mathcal{L}_{e_\perp} V^\perp + \frac{(\nabla_m N)}{N} V^m + A_\perp(V^\perp) \ , \tag{5a}$$

$$\mathbb{D}_{e_\perp}\mathbb{V}^b = \mathcal{L}_{e_\perp} V^b + \frac{\nabla^b N}{N} V^\perp + \frac{1}{2}\gamma^{bn}(\mathcal{L}_{e_\perp}\gamma_{nm})V^m + A_\perp(V^m) \ , \tag{5b}$$

$$\mathbb{D}_{\partial_a}\mathbb{V}^\perp = D_{\partial_a} V^\perp + \frac{1}{2}(\mathcal{L}_{e_\perp}\gamma_{am})V^m \ , \tag{5c}$$

and

$$\mathbb{D}_{\partial_a}\mathbb{V}^b = D_{\partial_a}V^b + \frac{1}{2}\gamma^{bm}(\mathcal{L}_{e_\perp}\gamma_{ma})V^\perp \quad . \tag{5d}$$

Applying this decomposition to the formulae for the curvatures, one obtains expressions for $\mathbb{F}^B{}_{\mu\nu}$ and $\mathbb{R}^\alpha{}_{\beta\mu\nu}$ entirely in terms of the spatial fields, their derivatives, and the intrinsic curvatures ($F^B{}_{ab}$ and $R^a{}_{bmn}$):

$$\mathbb{R}^\perp{}_{d\perp a} = \frac{1}{2}\mathcal{L}_{e_\perp}\mathcal{L}_{e_\perp}\gamma_{da} - \frac{1}{4}\gamma^{mn}\mathcal{L}_{e_\perp}\gamma_{dm}\mathcal{L}_{e_\perp}\gamma_{an} - \frac{1}{N}\nabla_d\nabla_a N \quad , \tag{6a}$$

$$\mathbb{R}^\perp{}_{cab} = \frac{1}{2}\nabla_a\mathcal{L}_{e_\perp}\gamma_{bc} - \frac{1}{2}\nabla_b\mathcal{L}_{e_\perp}\gamma_{ac} \quad , \tag{6b}$$

$$\mathbb{R}^d{}_{cab} = R^d{}_{cab} + \frac{1}{4}\gamma^{dm}(\mathcal{L}_{e_\perp}\gamma_{ma}\mathcal{L}_{e_\perp}\gamma_{bc} - \mathcal{L}_{e_\perp}\gamma_{mb}\mathcal{L}_{e_\perp}\gamma_{ac}) \quad , \tag{6c}$$

$$\mathbb{F}^B{}_{\perp b} = \mathcal{L}_{e_\perp}A^B{}_b - \frac{\nabla_b N}{N}A^B{}_\perp - D_b A^B{}_\perp \tag{7a}$$

and

$$\mathbb{F}^B{}_{ab} = F^B{}_{ab} \quad . \tag{7b}$$

2) *First Order Form*: With the fields, the derivatives, and the curvatures in 3+1 form, one may proceed to obtain the constraints and evolution equations. This may be done either by working with the spacetime action or the spacetime field equations. Here, we choose to work with the latter.

Substituting (5),(6) and (7) into (1) and (2), one finds that the equations are second order in time-derivatives. It is useful to reduce to first order. The most useful substitutions to use for this reduction are

$$K_{ab} := -\frac{1}{2}\mathcal{L}_{e_\perp}\gamma_{ab} \tag{8}$$

and

$$E^b{}_a := \mathcal{L}_{e_\perp}A^B{}_a - \frac{\nabla_a N}{N}A^B{}_\perp - D_a A^B{}_\perp \quad . \tag{9}$$

(We note that K_{ab} and $E^B{}_a$ are closely related to the canonical momenta for γ_{ab} and $A^B{}_a$).

3) *Primary Constraints*: In first order 3+1 form, 4+n of the field equations contain no time derivatives (n is the dimension of the group G). These are the primary constraints:

$$0 = \mathbb{D}_\mu \mathbb{F}^{D\mu\perp} = D_m E^{Dm} \quad , \tag{10}$$

$$0 = G_{\perp\perp} - (\mathbb{F}\,\mathbb{F})_{\perp\perp} = -R - (\mathrm{tr}k)^2 + K^m{}_n K^n{}_m + \frac{1}{2\lambda}E^B{}_m E^m{}_b + \frac{\lambda}{4}F^B{}_{aB}F^{ab}{}_B \quad , \tag{11}$$

and

$$0 = \mathbb{G}_{\perp b} - (\mathbb{F}\,\mathbb{F})_{\perp b} = 2(\nabla_m K^m{}_b - \nabla_b \mathrm{tr}K) + E_B{}^m F^B{}_{mb} \quad . \tag{12}$$

4) *Primary Evolution Equations*: The remaining 6+3n field equations all contain time derivatives and can be solved for them:

$$0 = \mathbb{D}_\mu \mathbb{F}^{B\mu b} \Rightarrow \mathcal{L}_{e_\perp} E^{Bb} = D_m F^{Bmb} + \frac{1}{2}(\mathrm{tr}k)E^{Bb} - E^b{}_F C^{FB}{}_M A^M_\perp + \frac{\nabla_m N}{N} F^{Bmb} \tag{13}$$

$$0 = G_{ab} - (\mathbb{F}\,\mathbb{F})_{ab} \Rightarrow \mathcal{L}_{e_\perp} K^a{}_b = R^a{}_b + (\mathrm{tr}K)K^a{}_b + \frac{\lambda}{2}[E^A{}_a E_{Ab} + F^A{}_{am} F_A{}^m{}_b]$$

$$- \frac{\lambda}{4}\delta^a{}_b [E^A{}_m E_A{}^m - \frac{1}{2} F^A{}_{mn} F_A{}^{mn}]$$

$$- \frac{1}{N}\nabla^a \nabla_b N \tag{14}$$

[In eq. 13, $C^{FB}{}_M$ are the structure constants of the group G.]

5) *Constraint Preservation*: The constraints and evolution equations obtained above are called "primary" because, in principle, there could be more. These arise if the time derivatives of the constraints do not vanish. That is, for every constraint $C[\gamma,K,A,E,\text{etc.}] = 0$, one must calculate $\mathcal{L}_{e_\perp} C$ and set it to zero. The result of this demand could be any of the following: i) Automatic - If $\mathcal{L}_{e_\perp} C$ vanishes when the primary evolution equations and constraints are substituted in, then there are no new constraints or evolution equations and the theory is consistent. ii) New Evolution Eq. - If $\mathcal{L}_{e_\perp} C$ may be solved for the time derivative of a variable which does not yet have a primary evolution equation, then one obtains a new evolution equation. The theory is still consistent. iii) New Constraint: If $\mathcal{L}_{e_\perp} C$ contains no time derivatives (after using the evolution equations) and doesn't vanish automatically, one obtains a new constraint. If the constraint is insoluble (e.g., 1 = 0) then the theory is inconsistent. Otherwise, the new constraint must be preserved. The same three cases can arise, and one proceeds until an inconsistency is reached or until all constraints <u>are</u> preserved (in which case the theory is consistent). In the case of the EYM theory, one finds that all 4+n constraints are preserved automatically.

6) *Identification of Gauge Evolvers* - Some of the fields - namely $A^B{}_\perp$, N and M^m - have no evolution equations. These are not dynamic variables, since their values in the future are <u>not</u> determined by their values (or the values of any other fields) in the past. In fact, their job is to control the evolution of the constraints and gauges. [See below.]

7) *Identification of Cauchy Data Fields* - The rest of the fields - namely γ_{ab}, $K^b{}_c$, $A^B{}_a$ and $E^B{}_m$ - do have evolution equations and therefore are determined by initial data. They constitute the Cauchy data for the EYM theory.

8) *Degrees of Freedom* - Since the Cauchy data are constrained and are neither gauge nor coordinate invariant, they are not all degrees of freedom. The number of degrees of freedom is given by

$$\text{DF's} = \frac{1}{2}\left[\binom{\text{pieces of}}{\text{Cauchy data}} - (\text{constraints}) - \binom{\text{gauge}}{\text{evolvers}}\right]$$

$$= \frac{1}{2}[(12+6n) - (4+n) - (4+n)] = 2+2n. \tag{15}$$

II. Building EYM Spacetime Solutions

We consider now how one uses the 3+1 formulation just obtained to build EYM spacetime solutions.

1) *Choose* S *and* B_S: Before one picks the initial data, one must make a choice of the initial surface S and the G-bundle B_S over it. These choices are important, since a spacetime solution built via 3+1 necessarily has for its spacetime manifold $M = \mathbb{R} \times S$, and B_M over M is necessarily the pullback of B_S. Note that these conditions strongly limit the variety of bundles one can obtain in 3+1 EYM solutions. If, for example, one chooses S to be the 3-plane $(\mathbb{R}^3)$ or the 3-sphere (S^3), then B_M must be trivial regardless of the group G. Or if one chooses G to be semisimple - SU(n), Sp(n), Spin (n), G_2, F_4, F_6, F_7 or F_8 - then again B_M must be trivial, regardless of the choice of the 3-surface S. Of course there are some choices of S and G which permit nontrivial bundles: A simple and perhaps physically interesting choice is $S = T^3$ (3-torus) and $G = SO(3)$.

2) *Choose the Initial Data*: The EYM initial data - a Riemannian metric γ_{ab} on S, a G-connection $A^B{}_b$ on B_S, a symmetric two-tensor K_{ab}, and a Lie algebra-valued vector $E^B{}_m$ - must be chosen to satisfy the constraints (13) and (14). These can be simplified by extending the York decomposition scheme[7]. That is, one finds that if the initial data is split as follows

$$\gamma_{ab} = \phi^4 \tilde{\gamma}_{ab} \quad , \quad K^b{}_d = \phi^{-6}[\tilde{\lambda}^b{}_d + (\tilde{L}W)^b{}_d] + \frac{1}{3}\delta^b{}_d \mathrm{tr}K,$$
$$A^B{}_m = \tilde{A}^B{}_m \quad , \quad E^m{}_B = \tilde{E}^m{}_b + \tilde{D}^m\theta_B \quad , \tag{16}$$

[with $\det \tilde{\gamma}_{ab} = 1$, $\nabla_m \mathrm{tr}K=0$, $\mathrm{tr}\,\tilde{\lambda}^b{}_d = 0$, $\tilde{\nabla}_m\tilde{\lambda}^m{}_b = 0$, $\tilde{D}_m\tilde{E}^m{}_B = 0$; and with $(\tilde{L}W)_{ab} := \tilde{\nabla}_a W_b + \tilde{\nabla}_b W_a - \frac{2}{3}\tilde{\gamma}_{ab}\tilde{\nabla}_m W^m$ while $\tilde{\nabla}_m\tilde{\gamma}_{ab} = 0$ defines $\tilde{\nabla}_m$][8], and if θ^B, W^m, and ϕ satisfy the elliptic equations

$$\tilde{D}_m\tilde{D}^m\theta^B = 0 \quad , \tag{17a}$$

$$\tilde{\nabla}_m(\tilde{L}W)^m{}_b = \tilde{F}^B{}_{mb}(\tilde{E}^m{}_B + \tilde{D}^m\theta_B) \quad , \tag{17b}$$

and

$$\tilde{\nabla}^2\phi = \frac{2}{3}K^2\phi^5 + \tilde{R}\phi - (\tilde{\lambda}^m{}_n + \tilde{L}W^m{}_n)(\tilde{\lambda}^n{}_m + \tilde{L}W^n{}_m)\phi^{-7}$$
$$- [\frac{1}{2}(\tilde{E}^m{}_b + \tilde{D}^m\theta_B)(E^B{}_m + \tilde{D}_m\theta^B) + \frac{1}{4}\tilde{F}_B{}^{mn}F^B{}_{mn}]\phi^{-3} \quad , \tag{17c}$$

then regardless of the choice of $\tilde{\gamma}_{ab}$, $\tilde{\lambda}^b{}_c$,trK,$\tilde{A}^B{}_m$ and $\tilde{E}^m{}_B$, (the "free data") the constraints (13) and (14) will be satisfied. Note that if solved in the proper order, (17a), (17b) and (17c) are decoupled. We also remark that for most choices of S, it has been shown that eqs (17) have unique solutions θ^B , W^m, and ϕ for (almost) all choices of the free data[7].

That a set of initial data satisfies the constraints is not enough to make it (and the solution constructed from it) physically interesting. The physics must be put in by hand. Fortunately, the free data is close enough to the physics to allow one to do this rather directly[9]. This is a distinct advantage of the 3+1 approach over other, less direct, procedures for obtaining solutions.

3) *Choose the Gauge Evolvers*: In solving the constraint equations (17) - indeed, in making sense of $\tilde{A}^B{}_m$ as a (Lie algebra-valued) one-form - one usually finds it necessary to make an explicit choice of the coordinate atlas on S as well as the local triviality atlas on B_S. So too if one is to have explicit formulae for the evolution of the initial data off the initial surface, one needs to know how these atlases change as one moves off the initial S and B_S to surfaces S_t and bundles B_{S_t} of the future[10].

The evolution equations, which are expressed most conveniently in the following form

$$\mathcal{L}_{\frac{\partial}{\partial t}} \gamma_{ab} = -2N\, K_{ab} + \mathcal{L}_{\vec{M}} \gamma_{ab} \ , \tag{18a}$$

$$\begin{aligned}\mathcal{L}_{\frac{\partial}{\partial t}} K^c{}_d = N[R^c{}_d + (\mathrm{tr}K)K^c{}_d + \tfrac{\lambda}{2}(E_A{}^c E^A{}_b + F_A{}^{cm} F^A{}_{mb}) \\ - \tfrac{\lambda}{4}\delta^c{}_d(E^A{}_m E_A{}^m - \tfrac{1}{2}F^A{}_{mn}F_A{}^{mn}) - \nabla^c\nabla_d N + \mathcal{L}_{\vec{M}} K^c{}_d \ ,\end{aligned} \tag{18b}$$

$$\mathcal{L}_{\frac{\partial}{\partial t}} A^B{}_m = N\, E^B{}_m + D_m(N\, A^B{}_\perp) + \mathcal{L}_{\vec{M}} A^B{}_m \ , \tag{18c}$$

and

$$\mathcal{L}_{\frac{\partial}{\partial t}} E_B{}^b = N[D_m F_B{}^{mb} + \tfrac{1}{2}(\mathrm{tr}k)E^{Bb} - E^b{}_F C^F{}_{BM} A_\perp{}^M] + (\nabla_m N)F_B{}^{mb} + \mathcal{L}_{\vec{M}} E_B{}^b \tag{18d}$$

appear to be explicit enough. But this is true only after N, $\vec{M}$ and $A^B{}_\perp$ have been specified on S along with the initial data. It is these three quantities which control the evolution of the atlases.

The appearance of each of the variables N, M^a and $A^B{}_\perp$ in the evolution equations (18) reflects what each controls:

N (the lapse), controls the proper distance from one t = const. surface to the next; i.e., it relates proper time to coordinate time (see fig.1). So in the evolution equation all terms except those involving $\mathcal{L}_{\vec{M}}$ are linear in N (or $\nabla_m N$).

$\vec{M}$ (the shift) controls the intrinsic motion of the spatial coordinates from one surface to the next (see fig.2). So there appears a Lie transport term along $\vec{M}$ in each evolution equation.

$A^B{}_\perp$ (the gauge shift) controls the intrinsic motion of the gauge coordinates from the bundle over one surface to that over the next. Since $A^B{}_m$ and $E^m{}_b$ depend upon the gauge coordinates while γ_{ab} and $K^c{}_d$ do not, $A^B{}_\perp$ appears in the evolution equations for the former two fields only. It contributes a gradient term to $\mathcal{L}_{\frac{\partial}{\partial t}} A^B{}_m$, as might be expected.

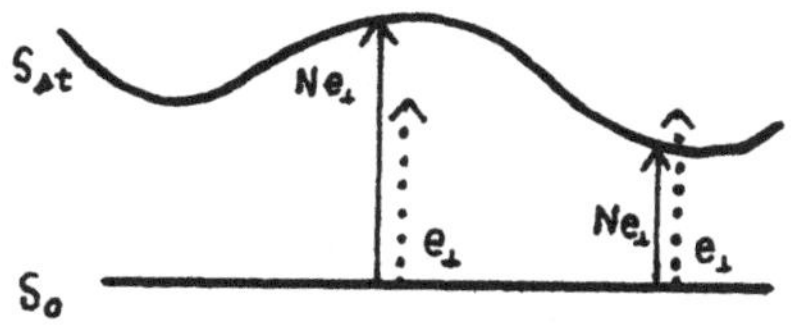

Fig.1 The lapse N determines the proper time between t=const. surfaces

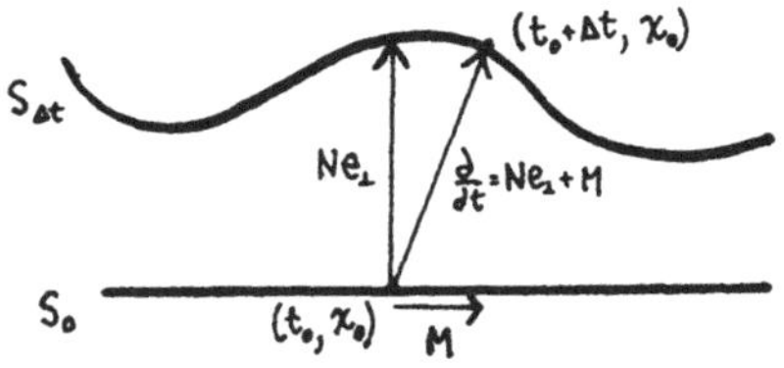

Fig.2 The shift $\vec{M}$ determines the motion of the coordinates

What values should one choose for these gauge evolvers on the initial surface (or any other)? Although the choice of N, $\vec{M}$ and $A^B{}_\perp$ cannot affect the physics of the solution being constructed, it certainly does affect the representation of that solution. Poorly chosen gauge evolvers can introduce spurious "coordinate waves" or "coordinate singularities" which obfuscate the physics and perhaps shut down the evolution calculation prematurely. Various rules for avoiding such problems have been found to be successful in a large variety of cases. (Choosing the lapse so that trK stays spatially constant is a fairly successful rule)[11]. But all of them have known failures. There is lots of room for research in this area.

By their basic nature, coordinates and gauges are of course patch-dependent. Thus the gauge evolvers can be chosen to be patch-dependent as well. Simple examples are known in which a patch-dependent choice of $\vec{M}$ and $A^B{}_\perp$ simplifies the represention of the EYM fields[12]. One must, however, be careful to properly evolve transition functions in accord with such patch-dependence. In particular, if on some patch-intersection $u_1 \cap u_2 \subseteq S$ one labels the coordinate and gauge transition functions by $\Lambda_{(12)}{}^{\bar{a}}{}_b$ and $\Psi_{(12)}{}^{\bar{B}}{}_E$, respectively, then one has

$$\mathcal{L}_{\frac{\partial}{\partial t}\big|_{u_2}} \Lambda_{(12)}{}^{\bar{a}}{}_b = \frac{\partial}{\partial x^b}\left(M^d|_{u_2} \Lambda_{(12)}{}^{\bar{a}}{}_d - M^{\bar{a}}|_{u_1}\right) \tag{19}$$

$$\mathcal{L}_{\frac{\partial}{\partial t}\big|_{u_2}} \Psi_{(12)}{}^{\bar{B}}{}_E = A_\perp{}^{\bar{B}}{}_{\bar{F}}|_{u_1} \Psi_{(12)E}^{\bar{F}} - \Psi_{(12)}{}^{\bar{B}}{}_F A_\perp{}^F{}_E|_{u_2} + \mathcal{L}_{\vec{M}|_{u_2}} \Psi_{12}{}^{\bar{B}}{}_E \quad , \tag{20}$$

(where $A_\perp{}^B{}_E : A_\perp{}^D C^B{}_{DE}$ is the adjoint representation of $A_\perp{}^D$, in the appropriate path). One also must be careful, when $\vec{M}|_{u_1} \neq \vec{M}|_{u_2}$, to prevent the surface from becoming uncovered. This is no problem as long as one extends the patch overlaps appropriately[13].

One might wonder at this point whether, as with the Einstein-Maxwell theory, one can replace the gauge-dependent variables {A,E} by gauge-invariant ones {B,E}, and thereby avoid worrying about gauge patching. Indeed, one can define a variable $B^B{}_a$ which is at least gauge-covariant and write out a pretty evolution system for it together with γ, K and $\mathbb{E}$. But the system is not complete without $A^B{}_m$ and $A^B{}_\perp$; hence it is not very useful.

4) *Evolve the data:* Evolving the Cauchy data from surface to surface using eqs. (18), (19) and (20) is a straightforward matter, regardless of how complicated the topology of S and B_S may be. Note that $\mathcal{L}_{\frac{\partial}{\partial t}}$ is the most useful time derivative operator since

it determines exactly the t-time change of a given quantity at a fixed spatial coordinate (evolving through time).

The computationally simple form of the evolution equations (suitable for numerical evolution by computer) belies the fact that one must prove that a given system such as this has unique solutions. For the EYM theory, such a proof has been given[14]. Therefore as long as one is careful in choosing the gauge evolvers on every surface S_t , one is guaranteed that one can construct a spacetime EYM solution from the chosen initial data.

5) *Reconstruct the Spacetime Solution:* The solution obtained is of course in the 3+1 form - a set of time dependent spatial fields on a set of copies of B_s over S. To convert to spacetime covariant form, one takes $M = \mathbb{R} \times S$, B = pull back of B_s from S to M, and then one reconstructs $\mathbf{g}_{\mu\nu}$ from (3) and $\mathbb{A}^B{}_\mu$ from (4).

We emphasize that the 3+1 procedure for constructing EYM solutions is quite practical (particularly with the numerical assistance of a computer) and should help us to understand the classical nature of the EYM theory (and possibly its quantum nature as well).

III. Degrees of a Freedom in a Theory with Dynamic Torsion

We proceed now to briefly discuss a very recent application of 3+1 analysis to the study of a field theory which is somewhat more complicated than EYM. This theory, proposed by Fairchild and by Yasskin as a possible "Yang-Mills theory for the Poincaré group", has been suspected to contain dynamic torsion fields (unlike the Einstein-Cartan theory in which the torsion is pointwise algebraically determined by the local spin density). The 3+1 analysis done by Yasskin and the author shows that this is the case.

The suspicion that the Fairchild-Yasskin ("FY") theory should have dynamic torsion stems from the fact that in the FY action

$$S_{FY} = \int d^4x \sqrt{-\mathbf{g}} \{ \mathbb{R} + \beta \mathbb{R}^{\alpha\beta\gamma\delta} \mathbb{R}_{\alpha\beta\gamma\delta} \} \tag{21}$$

there appear "$\partial_\mu \mathbf{\Gamma} \, \partial_\gamma \mathbf{\Gamma}$" type terms. However, while the absence of such terms (in the Einstein-Cartan action $S_{EC} = \int d^4x \sqrt{-\mathbf{g}}\, \mathbb{R}$, for example) guarantees that the torsion is not dynamic, the presence of them does not guarantee that all or part of it is dynamic. A more careful analysis, such as that sketched here, is needed.

The most convenient variables to work with in the FY theory are the tetrad frame $\mathbf{e}^\alpha$ and the connection $\mathbf{\Gamma}^\alpha{}_{\beta\mu}$. Varying them independently, one obtains the following spacetime covariant field equations:

$$0 = \mathbb{E}_{\alpha\beta} = \mathbb{G}_{\beta\alpha} + \beta(\mathbb{R}^{\kappa\lambda\mu}{}_\beta \mathbb{R}_{\kappa\lambda\mu\alpha} - \frac{1}{4} \eta_{\alpha\beta} \mathbb{R}^{\kappa\lambda\mu\nu} \mathbb{R}_{\kappa\lambda\mu\nu}) \tag{22}$$

and

$$0 = \mathbb{C}_{\beta\alpha}{}^\mu = \square_\nu (\boldsymbol{\delta}_{\hat\alpha}{}^{[\tilde\mu} \boldsymbol{\delta}_{\hat\beta}{}^{\tilde\nu]}) + \beta \square_\nu \mathbb{R}_{\hat\alpha\hat\beta}{}^{\widetilde{\mu\nu}} \quad . \tag{23}$$

[Here $\eta_{\alpha\beta}$:= diag (-1,1,1,1) are the tetrad components of the metric,and $\boldsymbol{\delta}_\alpha{}^\beta$ is the Kronecker delta. The index decorations are such that on an index $\hat\alpha$, one corrects with the full connection $\mathbf{\Gamma}^\alpha{}_{\beta\mu}$; while on an index $\tilde\mu$, one corrects with only the Christoffel part $\{^\alpha{}_{\beta\mu}\}$ of $\mathbf{\Gamma}^\alpha{}_{\beta\mu}$].

Let us now 3+1 . The tetrad frame decomposes into a spatial triad frame[15] θ^a along with the familiar lapse N and shift M^m. The spacetime connection decomposes into a spatial connection $\Gamma^a{}_{bc} := \mathbf{\Gamma}^a{}_{bc}$ as well as three other fields which we label as follows: $k_{ab} := \mathbf{\Gamma}^{\perp}{}_{ab}$, $\ell^a{}_b := \mathbf{\Gamma}^a{}_{b\perp}$, and $a^b := \mathbf{\Gamma}^b{}_{\perp\perp}$. When these variables, along with the properly decomposed covariant derivatives and curvatures are substituted into the field equations, one finds that the second derivatives of only $\Gamma^a{}_{bc}$ and k_{ab} appear. So only these variables require momenta, which we label $\rho_a{}^{bc}$ and π^{ab}, respectively.

One may now examine the field equations in first order (time derivative) form. All of the equations $0 = \mathbb{E}_{\alpha\beta}$ as well as $0 = \mathbb{C}_{\alpha\beta}{}^{\perp}$ turn out to be constraints. Although they are not all preserved automatically, the theory is consistent, as long as the secondary constraints $0 = \mathcal{L}_{e_\perp} \mathbb{E}_{a_\perp}$ and the secondary evolution equations $0 = \mathcal{L}_{e_\perp} \mathcal{L}_{e_\perp} \mathbb{E}_{a\perp}$ are satisfied. All field variables but N, M^a and $\ell^a{}_b$ have evolution equations, so one may summarize the 3+1 formulation of the FY theory as follows:

Dynamic Fields:	θ^a ,	k_{ab} ,	$\Gamma^a{}_{bc}$,	a^m ,	π^{ab} ,	$\rho_a{}^{bc}$	
	9	9	9	3	9	9	= 48
Gauge Evolvers:	N,	M^a ,	$\ell^a{}_b$				
	1	3	3				= 7

Constraints: $\mathbb{E}_{\perp\perp}, \mathbb{E}_{a\perp}, \mathbb{E}_{\perp a}, \mathbb{E}_{ab}, \mathcal{L}_{e_\perp}\mathbb{E}_{a\perp}, \mathbb{C}_{ab}{}^{\perp}, \mathbb{C}_{a\perp}{}^{\perp}$ = 25

Counting degrees of freedom, one finds $(48 - 7 - 25)/2 = 8$. Exactly how these 8 degrees of freedom fit in with the Poincaré group from the Yang-Mills perspective is not clear. There certainly is dynamic torsion (or, more properly, "dynamic connection") in the theory, however. Moreover, one should be able to find example solutions containing torsion (or connection) waves via 3+1 construction[16].

IV Spacelike Characteristics in a Generalized Gravito-Electromagnetic Theory

Another relatively complicated field theory for which 3+1 analysis has recently proven useful is the generalized gravito-electromagnetic theory, proposed by Horndeski. Using the same fields as the Einstein-Maxwell theory $\{\mathbb{G}_{\mu\nu}, \mathbb{A}_\mu\}$ but containing an extra term in its action,

$$S_H = \int d^4x \sqrt{-\mathbb{G}}\ \{ \mathbb{R} + \frac{1}{4}\mathbb{F}_{\mu\nu}\mathbb{F}^{\mu\nu} + \sigma\, \mathbb{F}_{\mu\nu}\, \epsilon^{\mu\nu\alpha\beta}\, \mathbb{R}_{\alpha\beta\gamma\delta}\, \epsilon^{\gamma\delta\lambda\xi}\, \mathbb{F}_{\lambda\xi} \} \quad , \tag{24}$$

(Here σ is a coupling constant) the H theory has been shown to be the most general theory which is (a) equivalent to Maxwell's theory in flat spacetime, (b) charge conservative, (c) derivable from an action principle, and (d) second order. The theory is therefore of some physical interest. It has, however the disturbing feature of spacelike characteristics. From the 3+1 perspective, these manifest themselves as follows: The evolution equations for the dynamic fields take the form

$$\begin{pmatrix} & M & \end{pmatrix} \begin{pmatrix} \dot{\gamma} \\ \dot{\pi} \\ \dot{A} \\ \dot{E} \end{pmatrix} = \begin{pmatrix} F \end{pmatrix}$$

where M is a matrix functional of the initial data, and F is a vector functional of the initial data as well as of the gauge evolvers. (Also, $\dot{\gamma}$, $\dot{\pi}$, etc, is shorthand for $\mathcal{L}_{\frac{\partial}{\partial t}} \gamma$, $\mathcal{L}_{\frac{\partial}{\partial t}} \pi$, etc.) For most values of the initial data, M is invertible and therefore the evolution of the initial data may be calculated. But there exist some choices of γ, π, A, and E such that M is degenerate. There are then more constraints and less evolution equations.

What happens to a solution when M goes degenerate? Do physical singularities appear? Might the data be discontinuous (a spacelike shock)? Is data for which M is degenerate avoided by a solution which starts at non-characteristic data?

To study these questions in the general class of solutions is difficult. Something can be learned, however, by looking at a restricted set. Let us, for example, consider the set of all solutions which are (1) spatially homogeneous, of Bianchi Type I ; (2) locally rotationally symmetric, and (3) purely electric. These conditions allow one to pick coordinates (t,x^a) and bases so that all of the dynamic fields $\{\gamma_{ab}, k^c{}_d, A_u, E^b\}$ depend upon t only, and so that $\gamma_{ab} = \mathrm{diag}(\gamma_1,\gamma_1,\gamma_3)$ $K^c{}_d = \mathrm{diag}\,(K^1,K^1,K^3)$, $A_a = (0,0,A_3)$, and $E^b = (0,0,E^3)$.

For fields of this form, the time development of γ_{ab} and of A_a are easily obtained as indefinite integral functions of $K^c{}_d$ and E^b, respectively. One may thus concentrate on the dynamics of K^1, K^3, and E^3. The picture further simplifies since the one nontrivial constraint[17] in the theory can be used to eliminate K^1. The entire system thus reduces to a two-dimensional dynamics problem.

The evolution equations for the two degrees of freedom are (using the variables $L := K^3$ and $E := \sqrt{\gamma_1}\, E^3$)

$$\frac{dL}{dt} = \frac{3}{2} L^2 - \frac{1}{2} E^2(1 - \sigma L^2) \tag{24}$$

and

$$(1 - \sigma L^2)\frac{d}{dt} E = EL[(2 - \sigma E^2) + \sigma L^2(1 + \sigma E^2)] \; . \tag{25}$$

One sees immediately that, even in the reduced system, the possibility for a degenerate M - for a spacelike characteristic - exists. It occurs when $L = 1/\sqrt{\sigma}$. So what happens at or near such data? The most illuminating picture of the behavior of the system is obtained from a phase diagram (Fig.3). One is immediately led to the following conclusions: (1) The characteristic data is avoided iff $E \neq 0$ (2) If $E = 0$, the solution passes right through the characteristic, with no discontinuities permitted. [This is the "LRS" Kasner solution.]

Further insight can be obtained by comparing the phase diagram for the Horndeski theory with that pertaining to the standard Einstein-Maxwell theory (assuming the same symmetries). One sees in Fig.4 that there are significant gross differences. However it is interesting to note that if one roughly locates "our universe today" on each phase diagram and compares the corresponding orbits, they are not very different.

So the effect of the spacelike characteristic data on these highly symmetric solutions is not as malevolent as it could be. Does the behavior get worse once the symmetry is removed? What happens in other theories? How (if at all) does this phenomenon affect the quantum picuture? All of these questions are important. The

3+1 analysis should be quite helpful in trying to answer them.

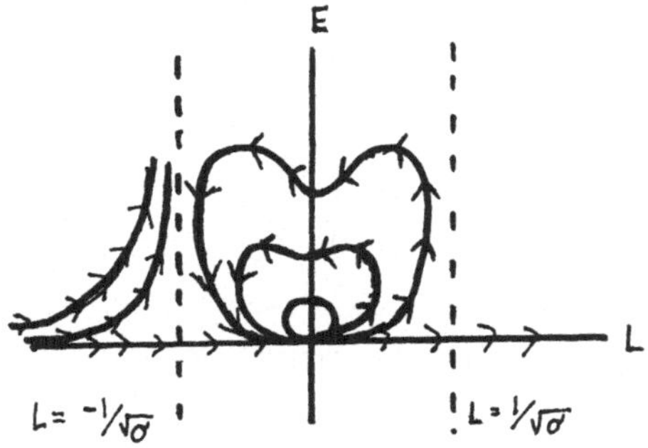

Fig.3 Phase diagram (E vs L) for Horndeski's Theory. The solid, arrowed lines represent solution trajectories. The dotted line indicates the characteristic data.

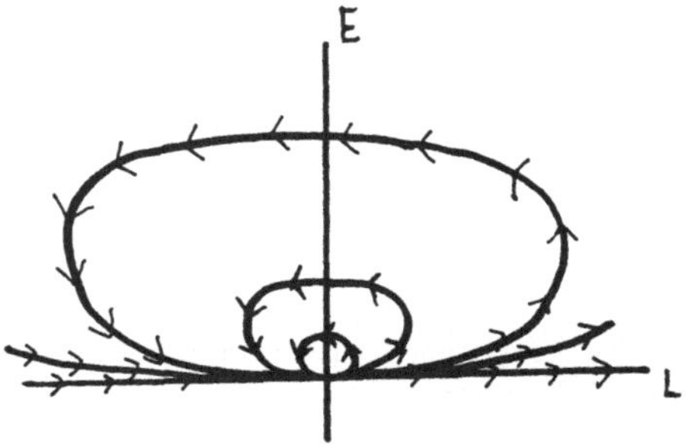

Fig.4 Phase diagram (E vs L) for Einstein-Maxwell Theory. The solid, arrowed, lines represent solution trajectories.

Footnotes

1. Note the increasing use of classical and semi-classical techniques in quantum field theory, as discussed in e.g. Coleman [1977], and Jackiw [1977].

2. See Choquet-Bruhat and York [1979], Fischer and Marsden [1979], Kuchar [1976], Isenberg and Nester [1979], and Isenberg [1979]. Our procedures and notation closely follow the latter two.

3. Our notation is as follows: Bold-faced tensors are spacetime covariant and regular-faced tensors are space covariant. The Riemannian and G-covariant derivative in spacetime is $\mathbb{D}_\mu$ (connection terms $\mathbf{\Gamma}$ and $\mathbb{A}$) and on the surfaces it is D_μ (connection terms Γ and A). The Riemannian covariant derivative on the surfaces is ∇_m (connection term Γ). Spacetime indices are greek, surface indices are lower case latin, and gauge indices are upper case latin. Our conventions on signs and signatures follow those of Misner, Thorne and Wheeler [1973].

4. See §2 and §3 of Isenberg and Nester [1979], and the references cited therein.

5. Comparing this with a coordinate basis, one has

$$e_\perp = \frac{1}{N}\left(\frac{\partial}{\partial t} - M^m \frac{\partial}{\partial x^m}\right) \qquad \theta^\perp = N\,dt$$

$$\frac{\partial}{\partial x^a} = \frac{\partial}{\partial x^a} \qquad \theta^a = dx^a + M^a\,dt \ .$$

6. Some subtleties arise in defining the proper Lie derivative for 3+1 evolution. (See §2.3 of Isenberg and Nester [1979]) but they can be ignored for present purposes.

7. See §3 of Choquet-Bruhat and York [1979] or chapter III of Isenberg [1979] for a general discussion of the York scheme. In III.C.3 of the latter reference, the York procedure is applied to the EYM theory.

8. Given all these conditions, it is a slight cheat to call this "free data". There are, however, canonical ways to generate the class of all transverse traceless tensors, for example. Note that the decomposition of $E^m{}_B$ into $\tilde{E}^m{}_B$ and $\tilde{D}^m\theta_B$ is really useful only when there are charged sources present [and then (17a) becomes $\tilde{D}_m \tilde{D}^m \theta^B = q^B$].

9. See III.D of Isenberg [1979].

10. S_t and B_s are a parametrized set of copies of S and B_{s_t} (with possible different coordinates and gauges).

11. See Marsden and Tipler [1979], Eardley and Smarr [1979], and chapter IV of Isenberg [1979].

12. For example, by properly choosing the shift (in a patch-dependent way) on a Kasner 3-torus cosmology, one can make the intrinsic geometry static.

13. See IV.G of Isenberg [1979].

14. The proof is a simple extension of the results on hyperbolic systems discussed in I-A of Choquet-Bruhat and York [1979] and in 3.4 of Fischer and Marsden [1979].

15. We require that the dual vector frame e_a be tangent to S.

16. The constraint problem is not as ugly as it appears. Many of the 25 constraints can be readily solved, algebraically, for certain of the "dynamic fields".

17. For general fields (i.e., no symmetry imposed),the Horndeski theory has the same number of constraints (and gauge evolvers) as does the Einstein-Maxwell theory.

References

Choquet-Bruhat, Y., and J. York, 1979, "The Cauchy Problem", in *Einstein Centenary Volume*, A. Held and P. Bergmann (eds.), Plenum, N.Y., (to be published).

Coleman, S., 1977, "Classical Lumps and Their Quantum Descendants", in *New Phenomena in Sub-Nulcear Physics*, A. Zichichi (ed.) Plenum, N.Y.

Eardley, D., and L. Smarr, 1979, Phys. Rev. D19, 2239.

Fairchild, E., 1977, Phys. Rev. D16, 2438.

Fischer, A., and J. Marsden, 1979, "The Initial Value Problem and the Dynamical Formulation of General Relativity" in *Einstein Commemorative Volume*, W. Israel and S. Hawking (eds.), Cambridge U. Press, Cambridge, U.K., (to be published).

Horndeski, G., 1976 J. Math. Phys. 17, 1980.

Isenberg, J., 1979, *The Construction of Spacetimes from Initial Data*, Ph.D. dissertatio University of Maryland (Physics), College Park, Md.

Isenberg, J., and J. Nester, 1979, "Canonical Gravity", in *Einstein Centenary Volume*, A. Held and P. Bergmann (eds.), Plenum, N.Y., (to be published).

Jackiw, R., 1977, Rev. Mod. Phys. 49, 681.

Kuchař, K., 1976, J. Math. Phys. 17, 777, 792, 801.

Marsden, J., and F. Tipler, 1979, "Maximal Hypersurfaces and Foliations of Constant Mean Curvature in General Relativity", Preprint, University of California at Berkeley (Math), Berkeley, Calif.

Misner, C., K. Thorne, and J. Wheeler, 1973, *Gravitation,* Freeman, San Francisco.

Yasskin, P., 1978, *Fibre Bundle Gauge Theories and Metric-Connection Theories of Gravity,* Ph.D. dissertation, University of Maryland (Physics), College Park, Md.

SOME SOLUTIONS OF THE EINSTEIN-YANG-MILLS EQUATIONS

Jacek Tafel[1]
Centre de recherche de mathematiques appliquees
Universite de Montreal
Montreal, Quebec, Canada

The geometrical approach to gauge field theory suggests some possibilities for finding new classical solutions of field equations. Recently Trautman [1] and Nowakowski and Trautman [2] noted that canonical connections on Hopf and Stiefel bundles automatically satisfy the source-free gauge field equations with respect to the natural metric structure on the base. A more general result, of which this may be seen as a consequence by projection, has been proved by Harnad and Shnider [3], namely[2]

Theorem Let (G,H) be a symmetric pair and $g = h+m$ the canonical decomposition of the Lie algebra of G. If m admits an $ad(H)$-invariant non-degenerate bilinear form B, then the bundle $G(G/H,H)$ with canonical connection and metric induced by B satisfies the source-free gauge field equations.

In general, the base manifold G/H is non-euclidean. The question arises whether the energy-momentum tensor of the gauge field corresponding to this canonical connection can be the source of the metric through the Einstein equations. In fact, it turns out that if G is semisimple this will always be the case but only with a suitable cosmological term. Since the cosmological constant seems to be physically undesirable it would be interesting to modify this result in some way in order to avoid it. A natural way is to identify the Riemannian symmetric space with the orbits under a G-action on a spacetime manifold which is not a homogeneous space. The simplest assumption is that the space decomposes into a product $V\times G/H$, where G acts only on the second term, and the bundle is identified as $P = V\times G$. We shall still take G to be semisimple. Let (t_α), (t_i) be bases in the subspaces h and m, respectively, of the algebra g. These are identified with left invariant vector fields on G. The dual set of 1-forms we denote by (θ^α,θ^i). These forms together with any co-frame θ^a on V can be considered as a co-frame on P. Define the connection on P as

$$\omega = \theta^\alpha t_\alpha .$$

Since G is semisimple there exists a natural G-invariant metric on G/H

---

[1] Permanent address: Institute of Theoretical Physics, Warsaw University, Hoza 69, Warsaw, Poland.

[2] We use notation based on [4].

$$d\ell^2 = K_{ij}\sigma_A^{*}\theta^i \otimes \sigma_A^{*}\theta^j,$$

where K_{ij} $(K_{\alpha\beta})$ is the restriction of the Killing form on g to m (h) and σ_A is any set of local sections $\sigma_A : U_A \to G$ defined on a covering (U_A) of G/H (the resulting metric being independent of the choice of σ_A). Consider metrics on V×G/H of the form

$$ds^2 = g_{ab}(V)\theta^a \otimes \theta^b + f^2(V)d\ell^2,$$

where g_{ab} and f depend only on coordinates on V. It can be easily verified that:

Proposition The connection ω satisfies the source-free gauge field equations on V×G/H with metric ds^2.

From the point of view of the gauge field, the introduction of the manifold V is rather trivial, but now a freedom of functions g_{ab} and f allows us to satisfy the Einstein equations without cosmological constant. For the right hand side of these we take the energy-momentum tensor of the gauge field derived from the action integral

$$\frac{\varkappa}{2}\int K_{\alpha\beta}\, \Omega^\alpha \wedge {*\Omega^\beta}$$

where $\Omega^\alpha t_\alpha$ is the curvature form of ω, star denotes dualization, and ϰ is a constant. Under this assumption the Einstein equations read as follows (in a system of units such that $8\pi G = 1 = c$):

$$R_{ab} - \frac{\bar{m}}{f}\nabla_a\nabla_b f + g_{ab}[-\frac{1}{2}R^c_c + \frac{\bar{m}}{f}\nabla_c\nabla^c f + \frac{\bar{m}(\bar{m}-1)}{2f^2}\nabla_c f\nabla^c f + \frac{\bar{m}}{4f^2}] = \frac{\varkappa\bar{m}}{8f^4}g_{ab},$$

$$[-\frac{1}{2}R^c_c + \frac{\bar{m}-1}{f}\nabla_c\nabla^c f + \frac{(\bar{m}-1)(\bar{m}-2)}{2f^2}\nabla_c f\nabla^c f + \frac{\bar{m}-2}{4f^2}]K_{ij} = \frac{\varkappa(\bar{m}-4)}{8f^4}K_{ij},$$

where all quantities are referred to the co-frame $(\theta^a, \sigma_A^{*}\theta^i)$, $\bar{m}$ is the dimension of G/H, and ∇_a and R_{ab} denote, respectively, the covariant derivative and the Ricci tensor on V corresponding to the metric tensor g_{ab}. The equations can be solved exactly when the dimension of V is one or two. For $f \neq$ const., the resulting metrics are as follows:

dim V = 1 $$ds^2 = dt^2 + [\frac{\varkappa}{2} - \frac{1}{2(\bar{m}-1)}t^2]d\ell^2$$

Assuming that V×G/H has dimension 4 (i.e. $\bar{m} = 3$) we obtain solutions very similar to the Robertson-Walker cosmological model [5] filled with electromagnetic radiation. For instance, when the symmetric space is $SO(4)/SO(3) \sim S_3$ (this bundle is trivial) then we have exactly

$$ds^2 = dt^2 - (2\varkappa - t^2)[d\chi^2 + \sin^2\chi(d\theta^2 + \sin^2\theta d\varphi^2)].$$

In this case the gauge field coincides with the Levi-Civita curvature of a 3-sphere [2].

dim V = 2 $$ds^2 = (1 + \frac{A}{r^2} - \frac{2M}{r^{\bar{m}-1}})dt^2 - \frac{dr^2}{(1 + \frac{A}{r^2} - \frac{2M}{r^{\bar{m}-1}})} + \frac{r^2}{2(\bar{m}-1)}d\ell^2$$

where

$$A = \begin{cases} \frac{(\bar{m}-1)^2}{(3-\bar{m})}\varkappa & \text{for } \bar{m} = 2 \text{ or } \bar{m} > 3 \\ -4\varkappa \ln r & \text{for } \bar{m} = 3 \end{cases}$$

and M is an arbitrary constant[3]. For $\bar{m} = 2$ the above solutions are the same type as the Reissner-Nordstrøm metric [5]. For instance, when G/H is $SU(2)/U(1) \sim S_2$ we have

$$ds^2 = (1 + \frac{\varkappa}{r^2} - \frac{2M}{r})dt^2 - \frac{dr^2}{(1 + \frac{\varkappa}{r^2} - \frac{2M}{r})} - r^2[d\theta^2 + \sin^2\theta d\varphi^2]$$

but the U(1) gauge field represents a magnetic monopole not an electric one.

The class of solutions presented in this paper is very restricted when a 4-dimensional base manifold is assumed. This is because of the isomorphisms between low dimensional Lie algebras. However higher dimensional models might be worthwhile considering provided imbeddings can be found into such higher dimensional manifolds which preserve the field equations. A class of such imbeddings was given in [1,2], but only with the gauge field equations considered.

The author wishes to acknowledge helpful discussions with J.Harnad, and is grateful for hospitality accorded during his stay at the C.R.M.A., where this work was done.

References

1. Trautman A., Intern.J.Theor.Phys. 16 (1977), 561.
2. Nowakowski J., Trautman A., J.Math.Phys. 19 (1977), 1100.
3. Harnad J., Shnider S., J. Tafel, preprint CRMA 922 (1979).
4. Kobayashi S., Nomizu K., Foundations of differential geometry (Wiley Interscience, 1969).
5. Misner C., Thorne K.S., Wheeler J.A., Gravitation (W.H.Freeman and Co., San Francisco)
6. Helgason S., Differential geometry and symmetric spaces (New York, Academic Press, 1962).

[3] In this case metrics satisfying conditions: $f^2 = \varkappa/2$, $R^a_a = -2/\varkappa$ are also allowed.

SOME INVARIANT SOLUTIONS TO THE YANG-MILLS EQUATIONS IN THE PRESENCE OF SCALAR FIELDS IN MINKOWSKI SPACE

Luc Vinet
C.R.M.A. and Département de Physique
Université de Montréal
C.P.6128, Montréal, Québec, Canada

1. Introduction

We would like to report here some solutions to the Yang-Mills equations coupled to an isotriplet of massless scalar fields in Minkowski space. Let A be the SU(2) gauge potential, that is a 1-form with values in the Lie algebra of SU(2), and let Φ be a scalar field with values in the adjoint representation of SU(2).

We shall take as basis for SU(2), the set $\{t_a=\sigma_a/2i,\ a=1,2,3\}$ where the σ_a are the usual Pauli matrices, and write

$$\Phi = \Phi^a t_a \quad (*) \tag{1}$$

The field equations we consider read as follows:

$$*D*DA = [\Phi, D\Phi] \tag{2}$$

$$*D*D\Phi = \lambda|\Phi^2|\Phi \quad (|\Phi|^2 = \Phi^a\Phi_a). \tag{3}$$

These equations are invariant under C(3,1), the conformal group of space-time provided we take

$$\Phi = \phi/(|g|)^{1/8} \tag{4}$$

where ϕ transforms as a scalar density with canonical scaling dimension $\ell = -1$. D denotes the covariant exterior derivative:

$$D = d + [A, \ ] \tag{5}$$

with gauge coupling constant e scaled to unity and * represents the Hodge (dual) operator. We shall be able to obtain solutions to the system (2)-(3) by demanding that the fields be invariant under some subgroups of the conformal group. Such an approach has already proved useful in finding solutions to the pure Yang-Mills equations in Minkowski space. See reference [1] for a review.

For C(3,1) to have a global realization, we must replace Minkowski space M by its conformal compactification $\bar{M}$ which is diffeomorphic to $S^1\times S^3$. We will thus write equations (2) and (3) on $S^1\times S^3$ and take as candidate solutions fields on $\bar{M}$ pertaining to given symmetry classes. Note that the immersion of M into $\bar{M}$ is performed by a conformal change of coordinates; accordingly since the equations we

(*) Summation over repeated indices is used throughout.

are considering are conformal invariant, any solution to these on $\bar{M}$ will remain a solution under its pull-back to M.

Harnad, Shnider and Vinet [2] (see also Forgács and Manton [3]) studied in details the conditions for a gauge field to be invariant under a group of space-time transformations and discussed the problem of constructing the most general gauge fields possessing a given symmetry. These questions as they have shown, are best dealt with in the framework of fiber bundle theory. We will briefly summarize some of their results to begin with and will formulate the invariance conditions for matter fields which were not yet discussed fully. The invariant fields on $\bar{M}$ to be used as Ansätze in eqs (2) and (3) will subsequently be given and we will then proceed in describing some of the symmetric solutions we were able to obtain in that way.

2. Invariance Conditions for Gauge and Matter Fields

Let $\{U_\alpha\}$ be an open covering of some differentiable manifold M (e.g. space-time). Denote by H the gauge group and by $\mathfrak{h}$ its Lie algebra. We take the view-point according to which a gauge potential is given as a connection form ω on a principal H-bundle P over M. In order to specify the potential in a local gauge one has to take a local section of the bundle:

$$\sigma_\alpha : U_\alpha \to P, \qquad \pi \circ \sigma_\alpha = \text{id}. \tag{6}$$

The physicist's potential is then defined as the pull-back of ω by σ_α:

$$A_\alpha = \sigma_\alpha^* \omega \tag{7}$$

and a change of gauge simply interpreted as a change of section.

Let G be a transformation group acting differentiably on M. To speak of invariance at the bundle level, one needs to specify the lifts to P of the G action on M. There are in general inequivalent possibilities and the problem arises of classifying principal H-bundles with G action projecting to that on M. In terms of coordinates a G action $\tilde{f}_g$ on P is determined by some functions $\rho_\alpha : U_\alpha \times G \to H$ acting as transition functions between σ_α and its image under G:

$$\tilde{f}_g \sigma_\alpha(x) = \sigma(f_g x) \rho_\alpha^{-1}(g,x). \tag{8}$$

We must have

$$\rho_\alpha(g_1 g_2, x) = \rho_\alpha(g_2, x)\, \rho_\alpha(g_1, g_2 x) \tag{9}$$

for the group composition law to be verified while these functions $\{\rho_\alpha\}$ should further satisfy certain compatibility conditions for the group action to be globally defined on P (see [2]). (We have assumed that the open sets U_α can be chosen G-invariant; if not, the elements of G appearing in (8) should be restricted to those for which $f_g x \in U_\alpha$.) Obviously to a different choice of section corresponds an equivalent set of functions ρ_α. The problem of classifying these sets was investigated in ref. [2].

Given such a G-action on P we may require that the connection form ω be invariant under it:

$$f_g^*\omega = \omega. \qquad (10)$$

This condition leads with the help of (8) to the following invariance condition for the local potential forms:

$$f_g^* A_\alpha(x) = \mathrm{Ad}\rho_\alpha(g,x)^{-1} A_\alpha(x) + \rho_\alpha(g,x)^{-1} d\rho_\alpha(g,x). \qquad (11)$$

Thus, in the base manifold, the invariance condition under a space-time transformation must also involve the gauge transformation functions ρ_α which define the group action on P.

The matter fields are interpreted as cross-sections of vector bundles E associated to P. Their symmetry properties under the transformation group G are thus closely linked to those of the gauge fields. Let V be the vector space isomorphic to each fiber of E and $D:H \to GL(V)$ be the representation through which E is associated to P. A cross-section of E, that is a map $\psi:M \to E$ such that $\pi_E \cdot \psi = \mathrm{id}$, can be characterized by a function $\tilde{\psi}:P \to V$ satisfying

$$\tilde{\psi}(ph) = D(h^{-1})\tilde{\psi}(p). \qquad (12)$$

Identifying thus the matter field with $\tilde{\psi}$, it is then attached to the base manifold M through the use of the section σ_α in the following way:

$$\psi_\alpha = \tilde{\psi}\circ\sigma_\alpha. \qquad (13)$$

Demanding that $\tilde{\psi}$ be invariant under a given G action like (8) on P:

$$f_g^*\tilde{\psi} = \tilde{\psi} \qquad (14)$$

we obtain the following invariance condition for ψ_α:

$$f_g^*\psi_\alpha(x) = D(\rho_\alpha^{-1}(g,x))\psi_\alpha(x). \qquad (15)$$

3. Conformal Transformations on Compactified Minkowski Space

Details concerning the compactification of Minkowski space and conformal group action may be found in ref. [4]. Here we shall summarize the necessary notation for the group actions we used.

Introducing the R^6 coordinates $\{u^\alpha, \alpha=0,1,\ldots,5\}$, a point in $\bar{M} \sim S^1 \times S^3$ is identified with the U(2) element

$$u = e^{-i\psi} v \qquad (16)$$

where

$$e^{i\psi} = u_5 + iu_0, \quad v = u^4 - iu^i\sigma_i \qquad (17)$$

with $u_0^2 + u_5^2 = 1 = u_1^2 + u_2^2 + u_3^2 + u_4^2$. The immersion $M \subsetneq \bar{M}$ gives the following correspondence with the Cartesian coordinates on M:

$$u^\mu = \pm \frac{x^\mu}{\tau}, \quad u^4 = \pm \frac{1+\underset{\sim}{x}^2}{2\tau}, \quad u^5 = \pm \frac{1-\underset{\sim}{x}^2}{2\tau} \qquad (18)$$

where

$$\tau = [x_o^2 + \tfrac{1}{4}(1-\chi^2)^2]^{1/2} \quad \text{and} \quad \chi^2 = x_o^2 - x_1^2 - x_2^2 - x_3^2. \tag{19}$$

We shall make use of the following conformal group actions on $\bar{M}$:

(i) Left translations under $SU(2)_L$:

$$L_w(u) = wu \qquad w \in SU(2) . \tag{20}$$

(For our purposes right translations lead to equivalent results.)

(ii) Left action of the product $SU(2)_L \otimes SU(2)_R$:

$$L_{w,w'}(u) = wuw'^{-1} \qquad w,w' \in SU(2). \tag{21}$$

This action is identifiable with that of the group SO(4) on the (u^1,u^2,u^3,u^4) coordinates.

(iii) Left or right translations under U(1):

$$L_\phi u = R_\phi u = e^{i\phi}u. \tag{22}$$

Again such an action is identifiable with that of SO(2) on the (u^o,u^5) coordinates.

A basis for the cotangent space $T^*[U(2)]$ is provided by the left-invariant canonical forms on U(2):

$$\omega_L^o = d\psi \tag{23a}$$

$$\omega_L^i = -2[\varepsilon^{ijk}u^j du^k + u^i du^4 - u^4 du^i] \tag{23b}$$

which satisfy the Maurer-Cartan structure equations:

$$d\omega_L^o = 0, \qquad d\omega_L^i + \tfrac{1}{2}\varepsilon^{ijk}\omega_L^j \wedge \omega_L^k = 0. \tag{24}$$

In equations (2) and (3), the * operator is defined with respect to the usual Minkowski metric $g_M = dx_o^2 - d\vec{x}^2$ which in the above notation is

$$\begin{aligned} g_M &= \tau^2(du_0^2 - du_1^2 - du_2^2 - du_3^2 - du_4^2 + du_5^2) \\ &= \tau^2[d\psi^2 - \tfrac{1}{4}\omega_L^i \otimes \omega_L^i]. \end{aligned} \tag{25}$$

The Minkowskian volume element is given by

$$dV = dx^o \wedge dx^1 \wedge dx^2 \wedge dx^3 = \frac{\tau^4}{8} d\psi \wedge \omega_L^1 \wedge \omega_L^2 \wedge \omega_L^3 . \tag{26}$$

4. Invariant Fields on $\bar{M}$

Let us now obtain the symmetric gauge and matter fields A and Φ (as defined in the introduction, H is now SU(2) and V = su(2)) that we will use as Ansätze in equations (2) and (3).

a) Invariance under $SU(2)_L$

Using the methods of refs. [2] and [4], the transformation function ρ_α may be reduced to the identity by a suitable choice of section σ_α.

Relative to the orthonormal frame $\{\tau d\psi, \frac{\tau}{2}\omega^i\}$ the determinant of g_M is τ^8, therefore eq. (4) becomes

$$\Phi = \frac{1}{\tau}\phi. \tag{27}$$

(Alternatively (27) may be interpreted as relating the density ϕ referred to an orthonormal basis where it equals Φ to ϕ.) Applying the invariance condition (15) to ϕ, we find the most general $SU(2)_L$ invariant field to be of the form

$$\phi = \begin{bmatrix} \phi_1(\psi) \\ \phi_2(\psi) \\ \phi_3(\psi) \end{bmatrix} \tag{28}$$

i.e. depending on the ψ-variable only.

For the gauge field A expressed in terms of the left-invariant canonical forms $\{d\psi, \omega_L^i\}$, applying eq. (11) yields the following expression for the most general $SU(2)_L$ invariant form

$$A = [A_j^i(\psi)\omega_L^j + B^i(\psi)d\psi]t_i. \tag{29}$$

The components A_j^i and B^i again depend only on ψ. Transforming to a cartesian basis $A = A_\mu^a dx^\mu t_a$ we find

$$A_o^i = \frac{1}{\tau^2}[-2A_j^i(\psi)x^o x^j + \frac{1}{2}B^i(\psi)(1+x_o^2+\vec{x}^2)] \tag{30a}$$

$$A_j^i = \frac{1}{\tau^2}[-2A_k^i(\psi)(\varepsilon^{jk\ell}x^\ell - x^j x^k - \frac{1}{2}(1+x^2)\delta^{jk}) - B^i(\psi)x^o x^j] \tag{30b}$$

where $\psi(x)$ is calculated from eqs. (17), (18).

b) Invariance under $SU(2)_L \otimes SU(2)_R$

Again, applying the methods of refs [2,4], one can show that there are two possible inequivalent choices of ρ when the gauge group is SU(2):

i) $\rho_\alpha[(g_L, g_R), x] = e$

ii) $\rho_\alpha[(g_L, g_R), x] = g_R$

In the first case, i.e. when $\rho = e$, invariance condition (11) leads to the following potential:

$$A = B^i(\psi)d\psi t_i \tag{31}$$

which gives a vanishing field; $F = DA = 0$. For the scalar fields, (15) gives rise again to (28) as the general expression.

In the second case, when $\rho[(g_L, g_R), x] = g_R$,

$$A = f(\psi)\omega_L^i t_i \tag{32}$$

is the most general invariant potential, and there is no invariant ϕ field except the trivial one $\phi = 0$.

There will be therefore no SO(4) invariant solution to our equations involving both non-trivial gauge and scalar fields. Nevertheless, one might hope to find solutions for SO(4)-invariant gauge potentials and $SU(2)_L$-invariant scalar fields. We come to that in the next section.

c) Invariance under U(1)

We can always add to the group actions we have just discussed a further U(1) action on S^1. Requiring this additional symmetry on the fields amounts to making the components A^i_j, B^i and ϕ^i constants.

5. The Equations for the $SU(2)_L$-Invariant Fields

We shall now use as Ansätze in equations (2) and (3) the invariant fields given in the last section. Since in all cases, these fields possess at least the $SU(2)_L$ symmetry, we shall first give the equations that result for the most general $SU(2)_L$ invariant fields. Inserting $\Phi = \frac{1}{\tau}\phi^i(\psi)t_i$ and A as given by eq. (25) in $*D*DA = [\Phi, D\Phi]$, we find the following reduced system of equations for the components $A^i_j(\psi)$, $B^i(\psi)$ and $\phi^i(\psi)$ of the fields on $S^1 \times S^3$:

$$\frac{d^2A^i_j}{d\psi^2} + \varepsilon_{imn}\frac{dB^m}{d\psi}A^n_j + 2\varepsilon_{imn}B^m\frac{dA^n_j}{d\psi} + 4A^i_j - 6\varepsilon_{imn}\varepsilon_{jpq}A^m_pA^n_q + 4(A^i_jA^m_\ell A^m_\ell - A^i_kA^m_jA^m_k)$$
$$+ (B^iB^mA^m_j - A^i_jB^mB^m) + (A^i_j\phi^\ell\phi^\ell - A^\ell_j\phi^\ell\phi^i) = 0 \qquad (33a)$$

$$4\varepsilon_{ijk}A^j_\ell\frac{dA^k_\ell}{d\psi} + 4B^iA^m_\ell A^m_\ell - 4A^i_\ell A^m_\ell B^m + \varepsilon_{ijk}\phi^j\frac{d\phi^k}{d\psi} + (B^i\phi^m\phi^m - \phi^iB^m\phi^m) = 0. \qquad (33b)$$

The same substitution in $*D*D\Phi = \lambda|\Phi|^2\Phi$ leads to

$$\frac{d^2\phi^i}{d\psi^2} + \phi^i + 2\varepsilon_{ijk}B^j\frac{d\phi^k}{d\psi} + \varepsilon_{ijk}\frac{dB^j}{d\psi}\phi^k + 4(A^k_\ell A^k_\ell\phi^i - A^i_\ell A^k_\ell\phi^k) + (B^iB^k\phi^k - B^mB^m\phi^i) + \lambda\phi^k\phi^k\phi^i$$
$$= 0. \qquad (34)$$

We shall now look for simultaneous solutions to eqs. (33) and (34). Note that these equations possess a residual invariance under gauge transformations preserving the $SU(2)_L$ symmetry, i.e. these depending only on ψ. The same equations are also invariant under the substitution

$$A^i_j \to A^i_jR^j_k, \quad B^i \to B^i, \quad \phi^i \to \phi^i \qquad (35)$$

as long as $R^j_iR^j_k = R^i_jR^k_j = \delta_{ik}$ as a result of the conformal invariance of the original equations. It follows that in general from a given solution, one can obtain an infinite class of non-equivalent solutions all related by $SU(2)_R$ translations. This will be taken as understood and not expressed explicitly in the resulting solutions. Now the higher the symmetry, the simpler is the Ansatz. Hence it seems reasonable to look for solutions with an O(4)-invariant gauge potential. The unique non-trivial case corresponds as we have seen to $A^i_j(\psi) = f(\psi)\delta^i_j$, $B^i = 0$. Unfortunately, even taking ϕ only $SU(2)_L$-invariant we find that for this Ansatz, equations (33) and (34) are not compatible unless either f or ϕ is identically zero, getting thus a pure gauge or pure scalar field theory. Schechter [5] has solved the equation resulting in the latter case while O(4)-invariant solutions to the pure Yang-Mills equations have been obtained independently by Lüscher [6] and Schecter [5].

If we are to find coupled solutions to our system we must relax some symmetry. Consequently we shall drop the $SU(2)_{R_1}$-invariance but in turn ask for additional invariance under the U(1) action on S^1. We thus search for constant solutions to (33) and (34) and from now on assume that the derivative terms in these equations vanish.

As a preliminary observation, note that if in eqs. (33) and (34), we set $\vec{B} = 0$ and make

$$\vec{\phi} \rightarrow i\vec{B} \tag{36}$$

the resulting equations are those obtained by setting $\vec{\phi} = 0$ in the same equations, provided

$$1 + \lambda\vec{\phi}^2 = 0. \tag{37}$$

In other words, if λ is picked to be $-1/\vec{\phi}^2$, the $SU(2)_L \otimes U(1)$ invariant solutions $\{\tilde{A}^i_j, \tilde{B}^i\}$ to the pure Yang-Mills equations (i.e. to eq. (33) with $\vec{\phi} = 0$ and no derivative terms) give solutions to the system (33)-(34) under the following identification:

$$A^i_j = \tilde{A}^i_j, \quad B^i = 0, \quad \phi^i = i\tilde{B}^i. \tag{38}$$

These pure Yang-Mills configurations are all known [4,7]. First there are two $O(4)\otimes O(2)$ invariant solutions [4-7] for which

$$\tilde{A}^i_j = f\delta^i_j \qquad \tilde{B}^i = 0 \tag{39}$$

with $f = 1$ or $\frac{1}{2}$.

The case $f = 1$ gives a pure gauge term while the case $f = \frac{1}{2}$ leads to the well known DeAlfaro, Fubini, Furlan solution [8]. There is another "diagonal" but complex solution of the form

$$\{\tilde{A}^i_j\} = \begin{bmatrix} 3 & & \\ & i & \\ & & i \end{bmatrix}, \quad \tilde{B}^i = 0. \tag{40}$$

Unfortunately, since $\tilde{B} = 0$, all these solutions lead to vanishing scalar fields. There is only one class of solutions for which $\vec{\tilde{B}}^2 \neq 0$, given by

$$\tilde{A}^i_j = \alpha^i\gamma_j, \quad \tilde{B}^i = \beta^i \tag{41}$$

where $\vec{\alpha}, \vec{\beta}$ and $\vec{\gamma}$ satisfy

$$(\vec{\alpha}\cdot\vec{\beta}) = 0, \quad \vec{\beta}^2 = 4 \quad \text{and} \quad \vec{\alpha}^2\vec{\gamma}^2 = 0. \tag{42}$$

Such solutions unless trivial must be complex; however they do provide a solution to eqs. (33)-(37) through the identification (38).

Obviously we want more interesting solutions. In particular, we would like to find real solutions (at least in their gauge part) to our equations. The next section is devoted to the further solutions we were able to obtain. We will describe all the $SU(2)_L \otimes U(1)$ invariant solutions for which the gauge field is of the form (41).

Next, we will determine a class of diagonal solutions extending solutions (39) and (40).

6. Some $SU(2)_L \otimes U(1)$ Invariant Solutions

We now present in summary form, solutions that belong to two subclasses of $SU(2)_L \otimes U(1)$ invariant fields. In Cartesian coordinates, the field F (the curvature 2-form) for a certain potential A is given by

$$F = DA = \frac{1}{2}\{F^a_{\mu\nu} dx^\mu \wedge dx^\nu\} t_a \tag{43}$$

where

$$F^a_{\mu\nu} = \partial_\mu A^a_\nu - \partial_\nu A^a_\mu + \varepsilon_{abc} A^b_\mu A^c_\nu . \tag{44}$$

It is convenient to define the analogues of electric and magnetic fields:

$$E^a_i = F^a_{io}, \quad H^a_i = \frac{1}{2}\varepsilon_{ijk} F^a_{jk}. \tag{45}$$

We shall also write

$$D\Phi = [(\mathcal{D}_\mu \Phi)^a dx^\mu] t_a = [(\partial_\mu \Phi^a + \varepsilon_{abc} A^b_\mu \Phi^c) dx^\mu] t_a . \tag{46}$$

In terms of these expressions the energy density for the system we are considering is given by

$$\theta_{oo} = \frac{1}{e^2}\{\frac{1}{2}((E^i_a)^2 + (B^i_a)^2 + [(\mathcal{D}_o \Phi)^a]^2 + [(\mathcal{D}_i \Phi)^a]^2) + \lambda|\Phi|^4\}. \tag{47}$$

θ_{oo} may then be integrated over any 3-dimensional space-like hypersurface to give the energy of the configuration.

We consider first solutions for which the gauge field takes the form:

$$A^i_j = \alpha^i \gamma_j \quad \text{and} \quad B^i = \beta^i . \tag{48}$$

Upon insertion of (48) into (33)-(34) we get the following system of algebraic equations,

$$4\vec{\alpha} + [\vec{\beta}(\vec{\alpha}\cdot\vec{\beta}) - \vec{\alpha}\vec{\beta}^2] + [\vec{\alpha}\vec{\phi}^2 - \vec{\phi}(\vec{\alpha}\cdot\vec{\phi})] = 0 \tag{49a}$$

$$4[\vec{\beta}\vec{\alpha}^2\vec{\gamma}^2 - \vec{\alpha}(\vec{\alpha}\cdot\vec{\beta})\vec{\gamma}^2] + [\vec{\beta}\vec{\phi}^2 - \vec{\phi}(\vec{\beta}\cdot\vec{\phi})] = 0 \tag{49b}$$

$$\vec{\phi} + 4[\vec{\phi}\vec{\alpha}^2\vec{\gamma}^2 - \vec{\alpha}(\vec{\alpha}\cdot\vec{\phi})\vec{\gamma}^2] + [\vec{\beta}(\vec{\beta}\cdot\vec{\phi}) - \vec{\phi}\vec{\beta}^2] + \lambda\vec{\phi}^2\vec{\phi} = 0 \tag{49c}$$

which we may solve completely. For all these configurations the energy is given by

$$E = \frac{2\pi^2}{e^2}\{4[\vec{\beta}^2\vec{\alpha}^2 - (\vec{\alpha}\cdot\vec{\beta})^2]\vec{\gamma}^2 + 16\vec{\alpha}^2\vec{\gamma}^2 + [\vec{\beta}^2\vec{\phi}^2 - (\vec{\beta}\cdot\vec{\phi})^2] + \vec{\phi}^2 + 4[\vec{\alpha}^2\vec{\phi}^2 - (\vec{\alpha}\cdot\vec{\phi})^2]\vec{\gamma}^2 + \frac{\lambda}{4}\vec{\phi}^4\}. \tag{50}$$

The solutions are:

1. $(\vec{\alpha}\cdot\vec{\beta}) = (\vec{\alpha}\cdot\vec{\phi}) = (\vec{\beta}\cdot\vec{\phi}) = 0$

   (a) $\vec{\beta} \equiv 0, \quad \vec{\phi}^2 = -4$

   $\vec{\alpha}^2\vec{\beta}^2 = \lambda - \frac{1}{4}, \quad E = \frac{8\pi^2}{e^2}(\lambda - 1).$ (51a)

(b) $\vec{\beta}^2 = \vec{\phi}^2+4$

$$\vec{\phi}^2 = 3/(\lambda-2) \qquad (\lambda\neq 2)$$
$$\vec{\alpha}^2\vec{\gamma}^2 = -(\vec{\phi}^2/4) \tag{51b}$$
$$E = \frac{18\pi^2}{e^2}\frac{(1-\frac{3}{4}\lambda)}{(2-\lambda)^2}$$

2. $(\vec{\alpha}\cdot\vec{\beta}) = (\vec{\alpha}\cdot\vec{\phi}) = 0, \qquad \vec{\beta} = \pm\frac{|\vec{\beta}|}{|\vec{\phi}|}\vec{\phi}$

$$\vec{\beta}^2 \equiv \vec{\phi}^2+4$$
$$\vec{\phi}^2 = -\frac{1}{\lambda} \qquad (\lambda\neq 0)$$
$$\vec{\alpha}^2\vec{\gamma}^2 = 0 \tag{52}$$
$$E = -\frac{3\pi^2}{2e^2}\left(\frac{1}{\lambda}\right).$$

3. $(\vec{\alpha}\cdot\vec{\beta}) = (\vec{\beta}\cdot\vec{\phi}) = 0, \qquad \vec{\alpha} = \pm\frac{|\vec{\alpha}|}{|\vec{\phi}|}\vec{\phi}$

$$\vec{\beta}^2 = 4$$
$$\vec{\alpha}^2\vec{\gamma}^2 = -\frac{\vec{\phi}^2}{4}$$
$$\vec{\phi}^2 = \frac{3}{\lambda} \tag{53}$$
$$E = -\frac{27}{2}\frac{\pi^2}{e^2}\left(\frac{1}{\lambda}\right).$$

We have obviously excluded the trivial solutions $\vec{\alpha} = 0$ or $\vec{\gamma} = 0$ leading to pure gauge fields as well as $\vec{\phi} = 0$. Note that solution 2. of which solution (42) described in the last section is a special case, always leads to a complex gauge field. If solutions 1. and 3. are to give real gauge fields, $\vec{\phi}$ must be pure imaginary while the range of λ should in each case be restricted appropriately.

The next set of solutions was obtained by setting a priori

$$\{A^i_j\} = \begin{bmatrix} a & & \\ & b & \\ & & b \end{bmatrix}, \quad \vec{B} = \begin{bmatrix} \beta \\ 0 \\ 0 \end{bmatrix}, \quad \vec{\phi} = \begin{bmatrix} \varphi \\ 0 \\ 0 \end{bmatrix}. \tag{54}$$

This Ansätz consists,in fact in the most general field configuration,left invariant not only by $SU(2)_L$ but also by those right $U(1)_R$ translations on S^3 which correspond to SO(2) rotations about the (t_1)-axis in the adjoint representation when the transformation function ρ is taken to be

$$\rho[(g_L,g_R),x] = g_R \qquad \forall\ g_R \in U(1)_R \subset SU(2)_R. \tag{55}$$

Consequently the solutions we will now describe possess a symmetry $SU(2)_L\otimes U(1)_R\otimes U(1)$ that is intermediate between $SU(2)_L\otimes U(1)$ and $O(4)\otimes O(2)$.

Substituting (54) into equations (33) and (34), we first find that $\vec{B}$ must vanish and obtain the following set of equations for a,b and φ:

$$a - 3b^2 + 2ab^2 = 0 \tag{56a}$$

$$1 - 3a + (a^2+b^2) + \frac{1}{4}\varphi^2 = 0 \tag{56b}$$

$$1 + 8b^2 + \lambda\varphi^2 = 0 \tag{56c}$$

The solution of this system is:

i) if $\lambda = 0$

$$a = -\frac{1}{2}, \quad b^2 = -\frac{1}{8}, \quad \varphi^2 = -\frac{21}{2} \tag{57}$$

φ is pure imaginary and the gauge field is complex.

ii) if $\lambda \neq 0$

(α) If $\lambda = 2$ the system (58) leads to

$$a^2 - 3a + \frac{7}{8} = 0 \tag{58a}$$

$$b^2 = -\frac{a}{2a-3} \tag{58b}$$

with solutions:

$$a_1 = \frac{3}{2} + \frac{1}{2}\sqrt{\frac{11}{2}}, \quad b_1 \simeq \pm i(1.07), \quad \varphi_1 \simeq \pm 2.01 \tag{59a}$$

$$a_2 = \frac{3}{2} - \frac{1}{2}\sqrt{\frac{11}{2}}, \quad b_2 \simeq 0.37, \quad \varphi_1 \simeq \pm i(2.14) \tag{59b}$$

Note that in the second case the gauge field is real.

(β) Finally if $\lambda \neq 2$, the system (56) is equivalent to the following:

$$(a-1)(2a-1)(a-3) + \frac{3}{4\lambda}(2a+1) = 0 \tag{60}$$

$$b^2 = (1-\frac{2}{\lambda})^{-1}[-a^2+3a+(\frac{1}{4\lambda}-1)] \tag{61}$$

$$\varphi^2 = \frac{1}{\lambda}(1+8b^2). \tag{62}$$

The solution thus involves solving a cubic equation for a; b^2 is directly given in terms of a and φ^2 in terms of b^2.

A full discussion of the actual solutions will be given elsewhere [9]. Let us make a few observations. Note first that if we let λ tend to infinity, φ vanishes and then, the solutions one gets from (60) and (61) are the three diagonal solutions (39) and (40) to the source-free Yang-Mills equations. Now since in general a cubic equation possesses three roots, the system (60)-(61)-(62) has a solution space with three branches, which may be connected by varying the value of λ. At the points $\lambda \to \pm\infty$, these three branches pass through the three source-free solutions.

These solutions may be interpreted therefore as interaction modes of the previously known source-free Yang-Mills configurations in the presence of an isotriplet of massless scalar fields. The solutions become complex for certain values of λ but in the range $-\infty < \lambda < -6.5$ the two branches of solutions connected to the pure gauge solution $a = b = 1$ and to the De Alfaro, Fubini, Furlan or meron solution $a = b = \frac{1}{2}$, are real in their gauge field and scalar field parts. At the point $\lambda = -6.5$ these two branches meet. On the other hand, for $\frac{1}{4} < \lambda < \infty$, the meron branch gives a real gauge field, while the scalar field is pure imaginary. The energy for this class of solutions may be obtained from the following formula:

$$E = \frac{\pi^2}{e^2}\{16[2b^2(a-1)^2+(-a+b^2)^2+\varphi^2[(1+8b^2)+\frac{\lambda}{2}\varphi^2]\}. \qquad (63)$$

Acknowledgements

I would like to thank J.Harnad for his constant help and suggestions during the course of this investigation. Very useful conversations with S.Shnider and P.Winternitz are also gratefully acknowledged. This work was supported in part by the Ministère de l'Education du Gouvernement du Québec.

References

[1] J.Harnad, S.Shnider and L.Vinet, in "*Complex Manifold Techniques in Theoretical Physics*", edited by D.Lerner and P.Sommers (Pitman Press 1979), pp.219-230.

[2] J.Harnad, S.Shnider and L.Vinet, "Group Actions on Principal Bundles and Invariance Conditions for Gauge Fields", CRMA preprint (1979), submitted to Comm.Math.Phys.

[3] P.Forgács and N.Manton, Comm.Math.Phys. (in press).

[4] J.Harnad, S.Shnider and L.Vinet, J.Math.Phys. 20, 931 (1979).

[5] B.Schechter,Phys.Rev. D16, 3015 (1977).

[6] M.Lüscher, Phys.Lett. 70B, 321 (1977).

[7] J.Harnad and L.Vinet, Phys.Lett. 76B, 589 (1978).

[8] V. de Alfaro, S.Fubini, G.Furlan, Phys.Lett. 65B, 163 (1976).

[9] L.Vinet, "Invariant solutions to the scalar-coupled Yang-Mills system in compactified Minkowski space" (in preparation).

Graded Riemannian Geometry and Graded Fibre Bundles, a Context for Local Super-Gauge Theories

Paul Green

Department of Mathematics, University of Maryland

Abstract

Starting from Kostant's definition of a graded manifold [1], we present graded versions of tensor algebra, Riemannian metrics, Lie groups and actions (these are also defined by Kostant but we present a directly geometrical definition which is more convenient for our purposes), vector bundles, and principal bundles.

With these notions in place, we can define a graded G-structure on a graded manifold. In the simplest non-trivial case, this leads immediately to a local version of the 14-dimensional graded Lie algebra discovered by Volkov and Akulov, and described by Ne'eman in [2].

Let M be a graded manifold (in the sense of Kostant) of even dimension 4 and odd dimension 4. We suppose that $T(M)$, the total graded tangent bundle of M, has a distinguished sub-bundle of even dimension 0 and odd dimension 4. We suppose further that the pairing $S \otimes S \to T(M)/S$ induced by the Lie-bracket induces a conformal class of Lorentz metrics on T/S by singling out the "squares" as null vectors. We observe that there is a complex structure on S which makes this Lie bracket pairing the real part of a Hermitian pairing from $S \otimes S \to T/S \otimes C$. Using this structure, it appears that one can give graded analogues of the Yang-Mills Ecuation and of the conformally invariant part of Einstein's equations.

References

[1] B. Kostant, Graded manifolds, graded Lie theory and prequantization, Differential Geometrical Methods in Mathematical Physics, Springer Lecture Notes in Mathematics #570 pp. 177-306.

[2] Y. Ne'eman, The application of graded Lie algebras to invariance considerations in particle physics, ibid pp. 109-144.

Group Actions on Principal Bundles and Invariance Conditions for Gauge Fields

J. Harnad, L. Vinet
Centre de Recherches Mathématiques Appliquées, Université de Montréal
S. Shnider
Department of Mathematics, McGill University

Abstract

Invariance conditions for gauge fields under smooth group actions are interpreted in terms of invariant connections on principal bundles. A classification of group actions as bundle automorphisms projecting to an action on the base manifold having a sufficiently regular orbit structure is given in terms of group homomorphisms and a generalization of Wang's theorem classifying invariant connections is derived. The strict invariance condition for a connection becomes invariance up to a gauge transformation for the local expression in a particular gauge. Criteria for the reduction of the auxiliary gauge transformation to simplest possible form are derived. The methods are demonstrated for several examples of gauge fields on compactified Minkowski space.

Details may be found in the references cited below.

REFERENCES

J. Harnad, S. Shnider, L. Vinet, "Invariance Conditions for Gauge Fields", Proceedings of the International Conference on Mathematical Physics, Lausanne, 1979. Springer Lecture Notes 1980, editor K. Osterwadler.

J. Harnad, S. Shnider, L. Vinet, "Group Actions on Principal Bundles and Invariance conditions for Gauge Fields, Preprint CRMA 899 (1979).

Algèbres de Lie d'ordre 0 sur une Variété

Pierre Lecomte
Institut de Mathématiques
Université de Liège

Abstrait

Un fibré vectoriel $E \xrightarrow{p} M$ dont la fibre type L est une algèbre de Lie est un fibré vectoriel en algèbres de Lie de type L s'il admet un cocycle à valeurs dans le groupe des automorphismes de L. Dans ces conditions, l'espace $\Gamma(E)$ des sections de classe C^∞ de E est muni d'une structure d'algèbre de Lie, appelée algèbre de Lie d'ordre 0 sur M .

Sous certains hypothèses plus restrictives sur L et sur M , on peut démontrer le théorème suivant : si E , E' sont deux fibrés en algèbres de Lie de type L et L' respectivement et si les algèbres de Lie $\Gamma(E)$ et $\Gamma(E')$ sont isomorphes, alors E et E' sont des fibrés isomorphes.

Ce théorème est une réponse partielle à une question de A.A. Kirillov concernant les algèbres de Lie locales. Il a de multiples applications. Par exemple la suivante: soient E , E' des fibrés vectoriels de base M et M' respectivement et soient $E_o(E)$, $[E_o(E')]$ l'algèbre de Lie des automorphismes infinitésimaux de E[E'] Si l'espace $H'(M, Z/2)$ de la cohomologie de Yech de M à valeurs dans les entiers module 2 est trivial, alors les fibrés vectoriels E et E' sont isomorphes si et seulement si les algèbres de Lie $E_o(E)$ et $E_o(E')$ sont isomorphes.

Parmis les exemples intéressants de fibrés en algèbres de Lie, le suivant est particulièrement important: soit P un fibré principal de groupe de structure G. Il est facile de montrer que le fibré associé à P pour l'action adjointe de G sur son algèbre de Lie G est un fibré vectoriel en algèbres de Lie de type G .

Les résultats ci-dessus sont l'objet d'une publication sous presse dans "Letters in Mathematical Physics" .

Homotopy Groups of the Space of Gauge Transformations

Chris Morgan
Memorial University of Newfoundland

Abstract

The following is joint work with Peter Booth, Renzo Piccinini and Philip Heath.

Let $p : E \to B$ be a principal G-bundle, $P_G : E_G \to B_G$ the universal G-bundle and $k : B \to B_G$ the classifying map for p. We denote by $G_1^{\sim}(p)$ the "group of gauge transformations" (i.e. the group of bundle automorphisms of p) and by $G_1^1(p)$ the group of based bundle automorphisms of p. We study the homotopy type and homotopy groups of these spaces.

If $L(B,B_G; k)$ denotes the component of the function space $L(B, B_G)$ of all maps $B \to B_G$ homotopic to k and if $L^*(B, B_G; k)$ denotes the component of the function space $L^*(B, B_G)$ of all based maps $B \to B_G$ homotopic to k, then $G_1^{\sim}(p) \simeq \Omega L(B,B_G; k)$ and $G_1^1(p) \simeq \Omega L^*(B,B_G; k)$. In particular, if B_{G_1} is an H-group, $G_1^{\sim}(p) \simeq L(B, G)$ and $G_1^1(p) \simeq L^*(B, G)$; if B is an H-co-group, $G_1^1(p) \simeq L^*(B, G)$.

Assume G is (n-1)-connected and $\dim B = m < 2n$. Then, for $0 \leq j \leq 2n - m - 1$, $\pi_j(G_1^1(p)) \simeq \pi_j(L^*(B, G))$ and $\pi_j(G_1^{\sim}(p)) \simeq \pi_j(L(B, G))$.

Connections on Infinitesimal Fibre Bundles and Unified Theories

D.K. Sen
Department of Mathematics
University of Toronto

Abstract

Mathematically, a gauge field is the curvature of a connection on a principal bundle over space-time with a certain structure group. Historically, Weyl [1] was the first to suggest such a connection with R^*, the multiplicature group of positive real numbers, as the structure group, in his pioneering attempt at a unified of gravitation and electromagnetism.

In trying to formulate intrinsically [8] another such unified theory, namely the projective theory of Jordan, Veblen et al [2], we [3] were led to the concept of an infinitesimal fibre bundle (M,ϕ,V,X,q), where

(i) M, V are smooth manifolds of dimensions $n+1$ and n, respectively,

(ii) ϕ, a submersion of M onto V.

(iii) X, a nowhere vanishing vector field on M such that if x,y lie on an integral curve of X then $\phi(x) = \phi(y)$, and

(iv) q is a connection 1- form on M, such that $q(X) = 1$ and the Lie derivative $L_X q = 0$.

Examples of infinitesimal fibre bundles are principal circle and line bundles (M,π,V,S'), (M,π,V, R). However, as we do not have the local triviality condition, the fibres $\phi^{-1}(v)$, at different points $v \in V$, of an infinitesimal fibre bundle are not diffeomorphic to each other - they are circles over some points and lines over others.

If now M is endowed with a metric suitable signature, there is a well-determined 2-form f as well as the curvature tensor on M, both of which can be projected onto V. It is then possible to identify V with space-time and $\phi_* f$ as the electromagnetic field tensor. Also the projection of the vacuum Einstein equations on M onto V give the Einstein-Maxwell equations.

It is possible to generalize the concept of infinitesimal fibre bundles where M and V are of dimensions n,m with $n = m+p$, $p=1,2,3,\ldots.$ It is suggested that this concept may provide a fruitful direction for the search of more comprehensive geometric structures for unified gauge theories.

References

[1] Folland, G.B. Journal of Diff. Geometry, 4. p. 145-53, 1970.

[2] Jordan, P. "Schwekraft und Weltall", Vieweg, Braunschweig, 1955, 2nd Ed.
Veblen, O. "Projective Relativitatstheoric", Springer, Berlin, 1933.

[3] Evans, G.T. & Sen, D.K., J. Math Physics, 14, no. 11, p. 1674, 1973.
Evans, G.T. Proc. Camb. Phil. Soc. 76, p. 465, 1974.

On Particles with Isotopic Spin

Jedrzej Sniatycki
Department of Mathematics and Statistics, The University of Calgary

Abstract
Classical dynamics of particles with internal degrees of freedom is described in terms of a principal fibre bundle over the cotangent bundle space of the space-time manifold. Geometric quantization associates, to each classical theory, a corresponding quantum theory. The internal degrees of freedom discussed here include charge, spin, and isotopic spin.

References

[1] Ch. Duval, "The general relativistic Dirac-Pauli particle: an underlying classical model" Ann. Inst. Henri Poincaré sec. Az vol. 25/1976/p. 345.

[2] V. Guillemin and S. Sternberg, "On the equations of motion of a classical particle in a Yang-Mills field and the principle of general covariance", Hadronic J. vol. 1/ 1978/p. 1.

[3] R. Kerner, "Generalization of the Kaluza-Klein theory for an arbitrary non-abelian gauge group", Ann. Inst. Henri Poincaré Sec. A. vol. 9/1968/p. 143.

[4] H.P. Kunzle, "Canonical dynamics of spinning particles in gravitational and electromagnetic fields", J. Math. Phys. Vol. 13/1972/p. 729.

[5] J. Sniatycki, "Geometric quantization and quantum mechanics", to appear.

[6] J.-M. Souriau, "Structure des systèmes dynamiques", Dunod, Paris, 1970.

[7] J.-M. Souriau, "Modele de particule a spin dans le champ electromagnetique et gravitationnel", Ann. Inst. Henri Pointcaré Sec. A. Vol.20/1974/p. 315.

[8] S. Sternberg, "Minimal coupling and the sympletic mechanics of a classical particle in the presence of a Yang-Mills field", Proc. Nat. Acad. Sci. U.S.A., Vol. 74/ 1977/p. 5253.

[9] A. Weinstein, "A universal phase space for particles in Yang-Mills fields", preprint.

Metric and Connection Theories of Gravity: The Gauge Theories of Spacetime Symmetry

Philip B. Yasskin*
Department of Physics, Harvard University

Abstract

A spacetime symmetry group is any group which may be used as the structure group for the tangent bundle to spacetime. I list many such groups and the corresponding principal bundles. The most familiar are: the Lorentz group = O(3,1,R) which uses the bundle of orthonormal frames, GL(4,R) with the general linear frame bundle, the Poincare group = IO(3,1,R) with the affine orthonormal frame bundle, and the spinor group, SL(2,C), with the orthonormal spinor frame bundle. The group GL(2,C) is also interesting because it leads to a unification of electromagnetism with a Weyl-Cartan theory of gravity.

Many authors have used these groups and bundles to describe gravity as a gauge theory. I summarize these efforts. As with all gauge theories, the principal bundle is given a connection which is taken as an independent variable. What distinguishes gravity from other gauge theories is the existence of a soldering form (a 1-form frame field dual to each vector frame field) which is also called the tetrad or vierbein. For the homogeneous groups (O(3,1,R), SL(2,C), etc.) the soldering form must be included as an independent variable in addition to the linear connection. For the inhomogeneous groups (IO(3,I,R), etc.) the soldering form may be identified as the translation part of the affine connection.

For those groups in which the connection is non-metric-compatible (GL(4,R) GL(2,C), etc.) the metric may be taken as an additional independent variable. In that case the metric acts as a Goldstone-Higgs field breaking the symmetry down to O(3,1,R) or SL(2,C) or IO(3,I,R) or ASL(2,C). The symmetry breaking is different from the usual Higgs mechanism because there are no residual massive scalar fields. If a kinetic term for the metric is included in the Langrangian then some components of the connection develop a mass.

* Supported by NASA grant NGR 21-002-010, and by a Chaim Weizman Fellowship

PARTICIPANTS IN SUMMER RESEARCH INSTITUTE

AXEL, F., CEN Saclay
BENCIVENGA, R., Memorial U., Newfoundland
BECKERS, C., Université de Liège
BERNARD, C., UCLA
BOOTH, P., Memorial U., Newfoundland
CORRIGAN, E., University of Durham
COUCH, E., Calgary University
FAIRLIE, D.B., University of Durham
FREUND, P.G.O., University of Chicago
GAUCHMAN, H., McGill University
GODDARD, P., Cambridge, U.K.
GREEN, P., University of Maryland
HARNAD, J., CRMA, Université de Montréal
HOFFMAN, P., University of Waterloo
HORNDENSKY, G.W., University of Waterloo
ISENBERG, J., University of Waterloo
JACKIW, R., MIT
KRAUSS, L., MIT
LAFRANCE, P., Université de Montréal
LECOMTE, P., Université de Liège
LYKKEN, J., MIT
McKELLAR, R.J., University of Waterloo
MOODY, G.P., Cambridge, U.K.
MORGAN, C., Memorial U., Newfoundland
O'RAIFEARTAIGH, L., Dublin
PATERA, J., CRMA Université de Montréal
PI, S-Y, SLAC
RAHA, S., University of Texas
REBBI, C., Brookhaven
ROSSI, P., MIT
RYMAN, A., University of Toronto
SEN, D.K., University of Toronto
ST-AUBIN, Y., Université de Montréal
SHARP, R., McGill University
SHNIDER, S., McGill University
SINHA, A., SUNY
SPARLING, G.A.J., University of Pittsburgh
SNIATYCKI, J., University of Calgary
TAFEL, J., Warsaw University
TORRENCE, R.J., University of Calgary
TRAUTMAN, A., Warsaw University
VINET, L., Université de Montréal
VOISIN, J., University of Burundi
WINTERNITZ, P., CRMA U. de Montréal
YASSKIN, P., Harvard

Selected Issues from

Lecture Notes in Mathematics

Vol. 662: Akin, The Metric Theory of Banach Manifolds. XIX, 306 pages. 1978.

Vol. 665: Journées d'Analyse Non Linéaire. Proceedings, 1977. Edité par P. Bénilan et J. Robert. VIII, 256 pages. 1978.

Vol. 667: J. Gilewicz, Approximants de Padé. XIV, 511 pages. 1978.

Vol. 668: The Structure of Attractors in Dynamical Systems. Proceedings, 1977. Edited by J. C. Martin, N. G. Markley and W. Perrizo. VI, 264 pages. 1978.

Vol. 675: J. Galambos and S. Kotz, Characterizations of Probability Distributions. VIII, 169 pages. 1978.

Vol. 676: Differential Geometrical Methods in Mathematical Physics II, Proceedings, 1977. Edited by K. Bleuler, H. R. Petry and A. Reetz. VI, 626 pages. 1978.

Vol. 678: D. Dacunha-Castelle, H. Heyer et B. Roynette. Ecole d'Eté de Probabilités de Saint-Flour. VII-1977. Edité par P. L. Hennequin. IX, 379 pages. 1978.

Vol. 679: Numerical Treatment of Differential Equations in Applications, Proceedings, 1977. Edited by R. Ansorge and W. Törnig. IX, 163 pages. 1978.

Vol. 681: Séminaire de Théorie du Potentiel Paris, No. 3, Directeurs: M. Brelot, G. Choquet et J. Deny. Rédacteurs: F. Hirsch et G. Mokobodzki. VII, 294 pages. 1978.

Vol. 682: G. D. James, The Representation Theory of the Symmetric Groups. V, 156 pages. 1978.

Vol. 684: E. E. Rosinger, Distributions and Nonlinear Partial Differential Equations. XI, 146 pages. 1978.

Vol. 690: W. J. J. Rey, Robust Statistical Methods. VI. 128 pages. 1978.

Vol. 691: G. Viennot, Algèbres de Lie Libres et Monoïdes Libres. III, 124 pages. 1978.

Vol. 693: Hilbert Space Operators, Proceedings, 1977. Edited by J. M. Bachar Jr. and D. W. Hadwin. VIII, 184 pages. 1978.

Vol. 696: P. J. Feinsilver, Special Functions, Probability Semigroups, and Hamiltonian Flows. VI, 112 pages. 1978.

Vol. 702: Yuri N. Bibikov, Local Theory of Nonlinear Analytic Ordinary Differential Equations. IX, 147 pages. 1979.

Vol. 704: Computing Methods in Applied Sciences and Engineering, 1977, I. Proceedings, 1977. Edited by R. Glowinski and J. L. Lions. VI, 391 pages. 1979.

Vol. 710: Séminaire Bourbaki vol. 1977/78, Exposés 507–524. IV, 328 pages. 1979.

Vol. 711: Asymptotic Analysis. Edited by F. Verhulst. V, 240 pages. 1979.

Vol. 712: Equations Différentielles et Systèmes de Pfaff dans le Champ Complexe. Edité par R. Gérard et J.-P. Ramis. V, 364 pages. 1979.

Vol. 716: M. A. Scheunert, The Theory of Lie Superalgebras. X, 271 pages. 1979.

Vol. 720: E. Dubinsky, The Structure of Nuclear Fréchet Spaces. V, 187 pages. 1979.

Vol. 724: D. Griffeath, Additive and Cancellative Interacting Particle Systems. V, 108 pages. 1979.

Vol. 725: Algèbres d'Opérateurs. Proceedings, 1978. Edité par P. de la Harpe. VII, 309 pages. 1979.

Vol. 726: Y.-C. Wong, Schwartz Spaces, Nuclear Spaces and Tensor Products. VI, 418 pages. 1979.

Vol. 727: Y. Saito, Spectral Representations for Schrödinger Operators With Long-Range Potentials. V, 149 pages. 1979.

Vol. 728: Non-Commutative Harmonic Analysis. Proceedings, 1978. Edited by J. Carmona and M. Vergne. V, 244 pages. 1979.

Vol. 729: Ergodic Theory. Proceedings 1978. Edited by M. Denker and K. Jacobs. XII, 209 pages. 1979.

Vol. 730: Functional Differential Equations and Approximation of Fixed Points. Proceedings, 1978. Edited by H.-O. Peitgen and H.-O. Walther. XV, 503 pages. 1979.

Vol. 731: Y. Nakagami and M. Takesaki, Duality for Crossed Products of von Neumann Algebras. IX, 139 pages. 1979.

Vol. 733: F. Bloom, Modern Differential Geometric Techniques in the Theory of Continuous Distributions of Dislocations. XII, 206 pages. 1979.

Vol. 735: B. Aupetit, Propriétés Spectrales des Algèbres de Banach. XII, 192 pages. 1979.

Vol. 738: P. E. Conner, Differentiable Periodic Maps. 2nd edition, IV, 181 pages. 1979.

Vol. 742: K. Clancey, Seminormal Operators. VII, 125 pages. 1979.

Vol. 755: Global Analysis. Proceedings, 1978. Edited by M. Grmela and J. E. Marsden. VII, 377 pages. 1979.

Vol. 756: H. O. Cordes, Elliptic Pseudo-Differential Operators – An Abstract Theory. IX, 331 pages. 1979.

Vol. 760: H.-O. Georgii, Canonical Gibbs Measures. VIII, 190 pages. 1979.

Vol. 762: D. H. Sattinger, Group Theoretic Methods in Bifurcation Theory. V, 241 pages. 1979.

Vol. 765: Padé Approximation and its Applications. Proceedings, 1979. Edited by L. Wuytack. VI, 392 pages. 1979.

Vol. 766: T. tom Dieck, Transformation Groups and Representation Theory. VIII, 309 pages. 1979.

Vol. 771: Approximation Methods for Navier-Stokes Problems. Proceedings, 1979. Edited by R. Rautmann. XVI, 581 pages. 1980.

Vol. 773: Numerical Analysis. Proceedings, 1979. Edited by G. A. Watson. X, 184 pages. 1980.

Vol. 775: Geometric Methods in Mathematical Physics. Proceedings, 1979. Edited by G. Kaiser and J. E. Marsden. VII, 257 pages. 1980.

Lecture Notes in Physics

Vol. 91: Computing Methods in Applied Sciences and Engineering, 1977, II. Proceedings, 1977. Edited by R. Glowinski and J. L. Lions. VI, 359 pages. 1979.

Vol. 92: Nuclear Interactions. Proceedings, 1978. Edited by B. A. Robson. XXIV, 507 pages. 1979.

Vol. 93: Stochastic Behavior in Classical and Quantum Hamiltonian Systems. Proceedings, 1977. Edited by G. Casati and J. Ford. VI, 375 pages. 1979.

Vol. 94: Group Theoretical Methods in Physics. Proceedings, 1978. Edited by W. Beiglböck, A. Böhm and E. Takasugi. XIII, 540 pages. 1979.

Vol. 95: Quasi One-Dimensional Conductors I. Proceedings, 1978. Edited by S. Barišić, A. Bjeliš, J. R. Cooper and B. Leontič. X, 371 pages. 1979.

Vol. 96: Quasi One-Dimensional Conductors II. Proceedings 1978. Edited by S. Barišić, A. Bjeliš, J. R. Cooper and B. Leontić. XII, 461 pages. 1979.

Vol. 97: Hughston, Twistors and Particles. VIII, 153 pages. 1979.

Vol. 98: Nonlinear Problems in Theoretical Physics. Proceedings, 1978. Edited by A. F. Rañada. X, 216 pages. 1979.

Vol. 99: M. Drieschner, Voraussage – Wahrscheinlichkeit – Objekt. XI, 308 Seiten. 1979.

Vol. 100: Einstein Symposion Berlin. Proceedings 1979. Edited by H. Nelkowski et al. VIII, 550 pages. 1979.

Vol. 101: A. Martin-Löf, Statistical Mechanics and the Foundations of Thermodynamics. V, 120 pages. 1979.

Vol. 102: H. Hora, Nonlinear Plasma Dynamics at Laser Irradiation. VIII, 242 pages. 1979.

Vol. 103: P. A. Martin, Modèles en Mécanique Statistique des Processus Irréversibles. IV, 134 pages. 1979.

Vol. 104: Dynamical Critical Phenomena and Related Topics. Proceedings, 1979. Edited by Ch. P. Enz. XII, 390 pages. 1979.

Vol. 105: Dynamics and Instability of Fluid Interfaces. Proceedings, 1978. Edited by T. S. Sørensen. V, 315 pages. 1979.

Vol. 106: Feynman Path Integrals, Proceedings, 1978. Edited by S. Albeverio et al. XI, 451 pages. 1979.

Vol. 107: J. Kijowski, W. M. Tulczyjew, A Symplectic Framework for Field Theories. IV, 257 pages. 1979.

Vol. 108: Nuclear Physics with Electromagnetic Interactions. Proceedings, 1979. Edited by H. Arenhövel and D. Drechsel. IX, 509 pages. 1979.

Vol. 109: Physics of the Expanding Universe. Proceedings, 1978. Edited by M. Demiański. V, 210 pages. 1979.

Vol. 110: D. A. Park, Classical Dynamics and Its Quantum Analogues. VIII, 339 pages. 1979.

Vol. 111: H.-J. Schmidt, Axiomatic Characterization of Physical Geometry. V, 163 pages. 1979.

Vol. 112: Imaging Processes and Coherence in Physics. Proceedings, 1979. Edited by M. Schlenker et al. XIX, 577 pages. 1980.

Vol. 113: Recent Advances in the Quantum Theory of Polymers. Proceedings 1979. Edited by J.-M. André et al. V, 306 pages. 1980.

Vol. 114: Stellar Turbulence. Proceedings, 1979. Edited by D. F. Gray and J. L. Linsky. IX, 308 pages. 1980.

Vol. 115: Modern Trends in the Theory of Condensed Matter. Proceedings, 1979. Edited by A. Pekalski and J. A. Przystawa. IX, 597 pages. 1980.

Vol. 116: Mathematical Problems in Theoretical Physics. Proceedings, 1979. Edited by K. Osterwalder. VIII, 412 pages. 1980.

Vol. 117: Deep-Inelastic and Fusion Reactions with Heavy Ions. Proceedings, 1979. Edited by W. von Oertzen. XIII, 394 pages. 1980.

Vol. 118: Quantum Chromodynamics. Proceedings, 1979. Edited by J. L. Alonso and R. Tarrach. IX, 424 pages. 1980.

Vol. 119: Nuclear Spectroscopy. Proceedings, 1979. Edited by G. F. Bertsch and D. Kurath. VII, 250 pages. 1980.

Vol. 120: Nonlinear Evolution Equations and Dynamical Systems. Proceedings, 1979. Edited by M. Boiti, F. Pempinelli and G. Soliani. VI, 368 pages. 1980.

Vol. 121: F. W. Wiegel, Fluid Flow Through Porous Macromolecular Systems. V, 102 pages. 1980.

Vol. 122: New Developments in Semiconductor Physics. Proceedings, 1979. Edited by F. Beleznay et al. V, 276 pages. 1980.

Vol. 123: D. H. Mayer, The Ruelle-Araki Transfer Operator in Classical Statistical Mechanics. VIII, 154 pages. 1980.

Vol. 124: Gravitational Radiation, Collapsed Objects and Exact Solutions. Proceedings, 1979. Edited by C. Edwards. VI, 487 pages. 1980.

Vol. 125: Nonradial and Nonlinear Stellar Pulsation. Proceedings, 1980. Edited by H. A. Hill and W. A. Dziembowski. VIII, 497 pages. 1980.

Vol. 126: Complex Analysis, Microlocal Calculus and Relativistic Quantum Theory. Proceedings, 1979. Edited by D. Iagolnitzer. VIII, 502 pages. 1980.

Vol. 127: E. Sanchez-Palencia, Non-Homogeneous Media and Vibration Theory. IX, 398 pages. 1980.

Vol. 128: Neutron Spin Echo. Proceedings, 1979. Edited by F. Mezei. VI, 253 pages. 1980.

Vol. 129: Geometrical and Topological Methods in Gauge Theories. Proceedings, 1979. Edited by J. Harnad and S. Shnider. VIII, 155 pages. 1980.